# Contents

| | |
|---|---|
| The Royal Way | 5 |
| Memorize This | 11 |
| Easy Does It. Not. | 19 |
| Embracing Difficulty | 23 |
| It's Not in the Syllabus | 33 |
| Speedy Math | 39 |
| Man, Machine, or Both? | 45 |
| Keys to Confidence | 49 |
| Creativity in Problem Solving | 55 |

CONTENTS

Let's Get There Faster!     63

Independent Study     69

A Training Methodology     77

Mathematical Competitions in the US     93

Appendix - Syllabus for Competitive Mathematics     107

# A Parents' Guide to Competitive Mathematics

Cleo Borac

Goods of the Mind, LLC, 2016

This edition published in 2016 in the United States of America.

Editing and proofreading: David Borac, M.Mus.
Technical support: Andrei T. Borac, B.A., PBK

All rights reserved. Except as permitted under the United States Copyright Act, no part of this publication may be reproduced or distributed in any form or by any means, or stored in a database or retrieval system, without prior written permission from the publisher, unless otherwise indicated.

Copyright Goods of the Mind, LLC.

Send all inquiries to:

Goods of the Mind, LLC
1138 Grand Teton Dr.
Pacifica
CA, 94044

A Parents' Guide to Competitive Mathematics
$1^{st}$ edition

# The Royal Way

*Non est regia ad Geometriam via.*

– Euclid

*There is no royal way to geometry.* The great Greek mathematician Euclid is believed to have said these words when his sponsor Ptolemy, the king of Egypt, complained that the text of the famous *Elements* was difficult and asked for shortcut to its meaning. In today's corporate parlance, Ptolemy's request is 'summarize, summarize,' while Euclid's reply uses the Murphyan philosophy 'if you can't convince them, confuse them.' In his famous response, Euclid suggests to the king that the ability to be humble is the mark of a good scientist. The difficulty of mathematics does not know of king and vassal: it is the same for everyone.

The purpose of this essay is to find a *royal way* not to the comprehension of mathematics, but to its learning. Somehow, the sheer difficulty of mathematics has always supported the Ptolemies of the world who would like to understand it without having to struggle. Over time, attempts to do better than Euclid and supply students with 'a simpler way to learn mathematics' have resulted in a jungle of materials, syllabi, and 'standards' that have managed to obscure not only the matter to be learned, but also the goal of learning.

In our desire to make the study of mathematics a painless and safe experience, we have managed to remove, to a large extent, the very purpose of learning mathematics: that of becoming creative intellectuals who are capable of building new knowledge. Creative individuals still exist, of course, but it increasingly looks like they exist thanks to their own drive and self-study habits rather than as a result of improvements in math education.

Too often, we simplify our goals by decreeing that mathematics is about numbers: counting, adding, multiplying. Other times, we say that mathematics is a high level game that can be played with tools ranging from calculators to symbolic computation software. At one end there are the parents whose effort is directed at making sure the student is able to perform basic computations correctly and rapidly and at the other end there are the parents who insist that the student is motivated only by the big picture, not by the nitty-gritty that calculators can do. And so the pendulum of opinions swings and, with it, the answer to the question 'What math should we teach our children?' The more we need mathematics, the more we seem to forget what it is about. *Mathematics is about creativity, abstraction, and generalization.* The fundamental question we should be asking is not 'what' but 'how': '*How* should we teach mathematics to our children?'

The syllabus we teach is, perhaps, not as important as *how* it is taught. The following anecdote shows how traditional materials and methods that we use can have 'side-effects.'

Anika was at the end of her first grade and had developed a passion for mathematics. In parallel to her school, she had studied Singapore Math and Beast Academy with her Ph. D. father. She now felt ready to add a new dimension to her abilities. The first question I asked Anika was:

'George had 28 little cars and he lost all of them except 9. How many little cars does George have now?'

And Anika replied: '19'.

'Unfortunately, that is not correct. How about reading the problem again?'

Anika re-did her computation thinking that maybe, unlikely as it seemed to her, she had made a mistake when borrowing for subtraction. When she obtained 19 again, I asked her to read the problem one more time, this time really slowly:

'Ah, I see. I read the problem too fast. The answer is 9!'

'No, Anika. You did not read the problem too fast. You actually did not read the problem. You scanned for the numbers in the problem and read only the question. Based on the question, you decided the problem was about subtraction. Anika, when you are given a page of problems that are meant to train you to add and subtract they all read like this, right?'

'Jim had 15 stamps and Tom had 11 stamps. How many stamps do they have together?'

or

'Lara has 9 dolls and Stella has 12 dolls. How many dolls do they have together?'

'After a couple of pages of similar problem statements you no longer care about who had what or what they had. Your brain has learned that the goal of the statements is for you to perform addition or subtraction. The story told by the statement is boring and has no relevance. Then, your process for reading the statement becomes: "scan for the numbers, read the question, add or subtract, done!" It no longer involves reading all the information provided.

The problems model *identical processes* and have thus trained you to take a shortcut: ignore the process, it's the same. This shortcut is clearly a hindrance if the problem statements *model a new, different process* from problem to problem. In this case, the statement *is relevant* and has to be carefully read.'

Anika's training so far, though diligent, had caused her to disregard the statement. From now on, I would have liked Anika to acknowledge that each problem has a story which may, or may not, be different from the story of the previous problem. She would have to read the statement carefully and imagine what is happening in it. I told her she would have to think of the problem as an adventure:

'You are the adventurer, discovering new territories. Before experiencing how amazing these territories are, you are worried that dangers may lurk. This is normal, since you have never been there before, so it's all right to worry a little. You have to live with the fact that *you may make a bad decision.* That's okay. After all, each unsuccessful attempt is a learning opportunity. *You're not supposed to win this game, you're supposed to learn how to play it.*'

I suggested to her mother that Anika should think and decide if she wanted to switch to a type of problem solving that was not *as safe* as her previous experience. It was obvious that Anika liked math: she got the correct answers with unvarying promptitude. How would she like it if she didn't? Did Anika have a passion for math or a passion for excelling at math? Because a passion for *excelling* might prevent her from learning more math - the math of the dangerous kind, where one has to read the statement *carefully* and *imagine all the possibilities*. But it turned out Anika understood exactly what I meant and decided that I was showing her math as it really is. She asked to continue lessons.

In her wisdom, (the 'wisdom of a five year old child' as Douglas Adams would put it,) Anika realized that math is not supposed to be put in a format fit for kings, but it is the kings who must become laborers just like everyone else.

Anika's experience shows that there is learning that turns us into worse problem solvers by reducing our attention span, our focus, and our ingenuity. My conjecture is that it is not a question of *what we learn* but of *how we learn* it. Lately, some schools no longer teach students how to subtract multi-digit numbers with borrowing, how to multiply multi-digit numbers, or how to perform long division. Since calculators are readily available, they no longer consider these skills to be meaningful.

What does 'meaningful' even mean? How do we make a distinction between the *meaningful* learning goals in mathematics and the meaningless ones? I think *we should avoid equating 'meaningful' with 'useful.'* We should carefully avoid classifying mathematical skills and concepts as 'useful' or 'not useful'. History has shown that a use will be found for pretty much anything mathematics can produce - concepts have been known to become useful even hundreds of years after being developed! 'Meaningful' comprises not only those concepts and skills that can be put to use immediately, but also those that may prove to be useful at some future time. How do we make sure that what we learn is useful not only now but continues to be useful and relevant in the future? We can define *meaningful learning* as learning that results in becoming better equipped for solving problems creatively.

A simple example of useful mathematical concept is that of prime numbers. Widely used in computer science nowadays, and perceived as useful by everyone who needs a secure connection online, prime numbers were studied ever since Eratosthenes (276 BC - 195 BC) invented his famous sieve algorithm. For 2000 years, prime numbers were studied and number theory based on their study grew and grew, without any apparent useful application. Suddenly, in the $19^{th}$ century, cryptographers invented codes based on large prime numbers. Mathematics is like the genetic code: a lot of the information seems redundant until, given some conditions are applied, a seemingly dormant gene saves the species. One cannot trim the genetic code of its apparent redundancy and, similarly, one cannot plan the study of mathematics on the basis of the utility of each concept. Some of the most creative applications of mathematics in technology are based on concepts that had been buried in mathematical literature for centuries.

In other words, let us stop arguing whether it is useful to learn long division or not and focus rather on what clever questions we can ask for the student to learn long division in an *interesting* manner that stimulates a creative response. Therefore, let us focus on *meaningful* learning (learning that increases the capacity to solve) instead of on *useful* learning (learning that is needed for some short term goal.) Imagine mathematics as a large square and the student of it as a small figure attempting to walk all over the entire square. By the end of its lifetime, the figure would have barely moved a bit away from the corner. This is how large a field mathematics is. One cannot hope to learn it all. Luckily, mathematical concepts are very strongly connected by logic. It is therefore possible to cover a large portion of it by *learning key concepts* and *learning how to make connections between concepts*. In this way, which is half learning and half making up more mathematics as we go, we can control a much larger field of concepts than if we endeavored to 'learn it', i.e. memorize concepts and steps of how to apply them. Notice that exercising creativity is an integral part of learning mathematics. By attempting to teach mathematics as a series of concepts and their direct applications, today's schools are depriving the student of the most useful aspect of studying it: the fact that only humans can create math.

Training the mind to learn by solving problems for the purpose of increasing the creative problem solving ability results inherently in developing the capacity to learn quickly and thoroughly any new concept. We should not learn for the purpose of using the knowledge right away but for the purpose of training the mind. Such training helps us ask questions that advance our knowledge, learn how to research a topic, and how to connect new knowledge to existing knowledge. Most importantly, problem solving develops the capacity to connect logically and creatively the concepts learned. It also helps define the 'key' concepts better, as the most frequently traveled destinations on the map of concepts. Someone who can do all this can learn anything very efficiently and apply it to problems naturally.

# Memorize This

*Of what significance are the things you can forget. A little thought is sexton to all the world.*

– Henry David Thoreau, Journal

One day, the parents of a group I teach proposed that they should run a math club in parallel with my class. Most of them are engineers at major technology companies, so we decided it would be great to have students spend additional time solving problems. Supplied with copies of the most popular competitions, they started meeting on a weekly basis. After a while, I found out they were bringing to class long lists of facts for students to memorize. They were doing this in order to train for 'speedy math,' one of the typical contest stages in several math and science competitions. In Machiavellian spirit, they found it natural that 'the end justifies the means,' i.e. as long as a student can perform a computation rapidly, what does it matter that he can do so because he memorized a fact and not because he understands how that fact applies to the example at hand? Does he avoid stepping in mud because he has been told not to ever step in mud, or because he anticipates he will get dirty if he does so?

My point here is that instructors should not micro-manage the process of memorizing facts. Instructors are tempted to do so because they anticipate that, eventually, information should end up being memorized. However, this is not a good reason for asking students to memorize a fact *before* using it. This approach is contrary to how memorization happens naturally. The brain memorizes information as a result of experience, not in anticipation of it! Specifically for learning mathematics, students should become familiar with the problem, attempt to solve it using their current ability, build (or be led to build by leading questions from the instructor) a solution, attempt to improve that solution, arrive at the fact in question. After repeating this process in a variety of contexts, the brain will end up memorizing facts that it perceived as interesting. It is unnatural for the brain to memorize a list because it cannot associate it with any interesting experience.

The parents who ran the math club, all excellent computer engineers, proceeded to teach students in the same way they would teach a machine: first memorize the formula, then run the program that applies it - if any errors are noticed, just fix them. This is the de facto approach that schools take as well. However, the natural process of memorization happens as a result of *exposure to experiences that we find interesting and compelling.* Of course, each person may have a different perception of what constitutes an interesting experience, and therefore, standardizing the process of memorization is no simple task. If we want to help students memorize for the long term, then we should provide them with as much exposure as possible to interesting angles on the information we present. Exercises and problems should incorporate interesting and creative repetitions of the subject matter rather than direct, mechanical applications. By substituting rote memorization for memorization that occurs naturally through exposure to interesting experiences, we undermine the very root of creative learning: the link between cause (interesting experience) and effect (fact to be memorized.)

What does a 'speedy math' test tell about us? That we have strong computation skills. However, we have to ask how those skills were acquired, because after all, pocket calculators also have them. Machines that are capable of calculating do so because a step-based procedure has been previously stored in their memory. Though they are incapable of *creating* such a procedure or of *improving* it, machines can *use* it with unfailing accuracy and speed. On the other hand, a human intelligence can be trained to develop its number sense to the point at which it can *create and improve* procedures. This has to be the goal of training students! Human intelligence must be trained to be able to *create and improve*.

How do we know that a student has good number sense? We can, of course, ask some clever questions (i.e. problems) that require intelligent shortcuts in order to be answered quickly. This is the supposed usefulness of 'speedy math.' However, from the output alone, it is impossible to distinguish between the human who has memorized a lot of rules by rote and applies them blindly and the human who cleverly makes them up as he goes along - or, at the very least, fully understands how the rules work. For this reason, 'speedy math' is - and should be - only a component of more extensive testing. By making students memorize computation shortcuts for the specific purpose of acing the 'speedy math' component, parents are effectively treating them as artificial, not natural, intelligences. Attempts to cram to develop speed in computation are taking the *need* for such speed out of context. The requirement for speed in computation makes sense only in a larger context, a grand scheme for training math skills. In such a scheme, time pressure has the role of motivating students to perform computations creatively. Students should be trained to build and amplify their capacity to *find shortcuts*. After a longer training period, they will perform mental computation just as fast as the students who crammed rules, but they will know *why* the rules work, their observation skills will skyrocket, while their ability to predict the next steps of the solution will become a source of inspiration time after time.

The key to the previous paragraph is the duration of training. It is obvious that a human brain can be trained to mimic an artificial intelligence or even a machine-like execution process. This training is less complex than training the capacity to react creatively and thus requires less time. Because *results are seen faster,* parents tend to prefer more mechanical training. Parents are focused on short term, even immediate, results of a training process. It seems irrelevant to them how the student got an answer, provided he got the correct one. And the faster the student is able to produce this correct answer,

## MEMORIZE THIS

the better the process must be, at least according to a superficial observer. The creative approach takes longer to train and requires more thoughtful planning - to foster it, parents must learn to stagger their expectations over a longer period of time.

So, should students memorize facts or not? Again, this is not the question. In the long term, facts will end up getting memorized. The more meaningful question is: when should students be expected to have memorized a fact?

*Memorizing a fact should only then occur when the student* **has become creative** *in the specific area of knowledge and is able to re-create facts from their logical precursors.*

The process of learning a skill should not *start* by memorizing facts, rules, and procedures. The student must start by feeling the need for a shortcut. As a simple example, the student has to calculate $8756 \times 8756 - 8756 \times 8755$. He can do the computations using the order of operations (slow), or he can notice that the difference is 8756 (fast). If the operation is $3 \times 3 - 3 \times 2$, the student will not need a shortcut, as the work is simple enough. The student will feel a need to simplify if the numbers are large, the computation looks difficult, and the allocated time is stringent. These conditions will force a creative response and teach the student about the importance of finding a shortcut. At this stage, it is productive to memorize this type of shortcut - namely, that it is useful to find common factors before processing the operations.

The process of acquiring mathematical ability is like the process of acquiring power: seize the most crucial assets and the rest will follow. In learning mathematics, the student should memorize the most crucial facts and information and use creativity to derive other facts on the spot, as needed. It is the teacher's role to identify the core of facts that need to be memorized and to provide insight into the process of using imagination and logic to derive other facts. The more knowledge we can make up, the more powerful we are as mathematicians.

# MEMORIZE THIS

In mathematics, when memorizing information, we have to pay attention to a fact's *importance*. A fact is more important when more facts depend upon it. By memorizing this fact we can derive other, dependent facts, from scratch if need may be.

However, as much as we would like to argue that memorization should be minimized, we cannot free ourselves from its burden. Those who talk about 'grasping the big picture' or think that we don't need to memorize an algorithm because a calculator can execute it, are considerably in error. Even if the calculator can execute it, only a human can *modify, adapt, extrapolate, or generalize* it! And for these purposes, it is important that students be able to derive every nut and bolt of the algorithm.

Memorizing facts and algorithms should not be the *primary* way of learning math. However, as they continue to solve problems, mathematicians will memorize facts for a variety of reasons: they are interesting, they provide more efficient ways to build a proof, they connect in some way with another topic, or they are aesthetically pleasing. In practice, all good problem solvers end up memorizing lots of facts, lots of interesting numbers, and lots of procedures. At the end of the day, each of us has their own baggage of memorized facts.

Parents often ask questions such as: 'What is the list of formulas that must be memorized in order to do well on AMC 10?' If only it were so simple. There are such lists and, for reference, one is included in the Appendix, but memorizing these formulas does little to help students become better at solving problems.

One reason a formula is not, in itself, an asset for the student who has diligently memorized it, is that a formula is merely a *shortcut* provided for efficiency in calculating. For this shortcut to be possible, some conditions must be met: every formula has a range of applicability. For example, the Pythagorean theorem applies only to right angle triangles. It is not seldom that I see students applying it to triangles that do not have a right angle. The people who write problems are aware of this and cunningly imagine situations that are either partially or totally outside the range of applicability of some formula. Students

## MEMORIZE THIS

who know *the proof* of the formula are aware of the limitations and will be able to improvise accordingly, often making up a bit of theory that extends what they have already learned. Students who memorize the formula without having an idea of how the formula came to be often fall in the traps that are routinely set for them.

A student who memorizes a formula without having a good understanding of why and how it works will, at some point, apply the formula outside its range of applicability.

*Though memorizing formulas, theorems, and algorithms does occur and must occur, formulas must not be learned through a process of systematic memorization, but through a process of* **investigation** *and* **mindful repetition***.*

During this process of *investigation* the student must ask questions such as:

- What does the formula help solve?
- How was the formula derived from its logical precursors?
- What restrictions on the applicability of the formula have been inherited from its logical precursors?
- What restrictions on the applicability of the formula have resulted from the process of proving it?

By *mindful repetition* I mean the repeated use of the formula in contexts that differ *substantially*. These contexts must aid not only the retention of the formula but also the exploration of its range of applicability, its connection to other areas of mathematics, its computational complexity, or its scalability to more complex problems in the same family.

For example, students can learn the quadratic formula. Then, they can study the analysis of quadratic functions (without and with parameters) and apply the quadratic formula to a diversity of analysis

## MEMORIZE THIS

problems or to inequalities. They can apply the quadratic formula to a number of geometry and coordinate geometry problems. A number of word problems related to minimizing/maximizing a quantity can be modeled with quadratic functions. Also, quadratic function analysis is needed to solve certain Diophantine equations. As the student progresses through this list of topics an ever deeper understanding of the quadratic function has to develop that goes beyond simply finding the roots by using the quadratic formula. A good exposition of the analysis (convexity, discriminant, Viète's relations) is needed from the start, but students will not retain all the details before they have experienced them in problem solving.

The syllabus and learning materials (such as worksheets) should support the process of mindful repetition by purposefully bringing back into focus the same concept at different levels of difficulty and in connection with other concepts.

Remember: facts and formulas are important for faster execution, not for understanding a topic! First take the time and effort to understand, then check for understanding by making sure there is a creative response, and lastly, let memory take hold of that distilled product of the learning process: the shortcut.

*MEMORIZE THIS*

# Easy Does It. Not.

*Easy is evil and has infinite forms.*

– Blaise Pascal, Thoughts

The greatest difficulty that modern students encounter when studying mathematics is the inability to focus. After having finished an examination, many students give one or several of the following reasons for their inability to meet their own expectations:

- 'I made a couple of stupid mistakes.'

- 'The problems were much harder than in previous years.'

- 'I ran out of time.'

Of these, the first is the most common. I have heard it so often in fact, that, when Sunitha asked me for tips, the first one that came to my mind was: "Eliminate 'stupid errors.' Many students emerge from the exam room and say 'I made some stupid errors.' This means they would have gotten the correct answer had they not made an error due to lack of focus. There is only one time to stop making these errors and that is NOW. They have to learn how to focus." A month later, she wrote back: "Awesome tips, Cleo. I shared your advice with my

## EASY DOES IT. NOT.

kids. The guideline about getting it right the first time - in the context of eliminating silly errors - was so impactful."

So what are stupid errors about? Stupid errors are avoidable errors that happen mainly because the student is not able to focus (lack of attention) or is not able to maintain focus for the entire duration of the exam (lack of stamina).

Most often, errors will happen in two portions of an exam: at the beginning, where the easy questions are, and at the end, where the more difficult questions are. Though it is understandable that there should be more errors on the difficult part of a test, it is more puzzling when errors also riddle the easy questions. Such errors can be explained by the inability to focus from the very start of the test. It takes the student a few minutes to fully focus on the task and it is during those minutes that the errors happen. As a result, we often see exam papers that have errors at the start, then show a string of correct answers, and then deteriorate again towards the end.

Another pattern of focus deficiency is a pattern of correct and incorrect answers that are not correlated with difficulty or topic. Many students focus intermittently, with periods of lack of attention during which their responses are very superficial.

Andrew L. was a very astute student enrolled at Harker High School who came to me with a 670 point score on the SAT Math Subject. By getting help identifying his deficiencies in focusing, he improved to a perfect score of 800 in 2 lessons! Since then, a lot of students in similar situations experienced sudden improvement. Nowadays, I sum it all up in a pep-talk I give a few weeks before the exam. This is it:

*This is a room full of young, ambitious, diligent, astute students who want nothing better than to excel on this test. Yet, a few days from now, most of you will come out of the exam room and start asking for sympathy on account of the 'stupid errors' they have made. Why do you make these errors? For the most part, it is due to lack of focus, or intermittent focus. You think a bit, you phase out, you think a*

## EASY DOES IT. NOT.

bit, you phase out... Now, why is that and how hard is it to fix? It is, for the most part, because you are given lots of easy homework. Easy homework is not harmless. Easy homework is boring, as you well know: you start on the boring math, from time to time you check to see if you have any messages; the chemistry test is also upcoming so you have a chat open with a friend who needs help. You do a bit more boring math while listening to some music. The phone rings from time to time, you do a bit more boring math, Janine posts on Facebook about her new shoes, Alex sends an urgent update on the DECA project. Your work process is like flipping channels on TV! It's okay for some types of work, but not for math. Nor, for that matter, for playing the trombone or for playing basketball. Easy homework creates this type of need for ceaseless multitasking, without focusing deeply on any one task. If this behavior becomes a habit, it is harmful. So how do you fix this? As with every habit, there is only one way to shake it: going cold turkey. Just decide now that you can prove to yourself that you do have the ability to focus totally, deeply, and for a long time, on a difficult problem. That's why you hate difficult problems - you cannot find solutions because you do not focus long enough on solving. And yet, it's the difficult problems that are the cure. It's by setting the bar higher that you will learn how to focus and how to remain focused until an idea forms in your brain - the idea that will help figure out a solution. Take a difficult problem now and focus on it. Don't go anywhere, don't do anything, don't take a washroom break or raid the refrigerator, don't reply to that message, don't eat, don't chew, don't do anything but breathe and think until you find a solution. Your parents are wrong when they say: 'It's okay. You will do better next time. It's obvious that you know how to answer these questions but just, for some reason, you made a silly error.' It's not okay. Next time you will do the same. Year after year you will make silly errors and your parents will sympathize. They have the feeling that you have to be rewarded for trying and you should not be discouraged from participating again. But no, remember the Latin saying: Errare humanum est, perseverare autem diabolicum[1]. You have to talk about the errors and decide to not excuse them. Unfortunately, there is no eliminating the tons of

---

[1]To err is human, to persist in such error is diabolical. (attributed to Seneca the Younger)

*easy homework, but there are still a few good things left: the difficult, frustrating problems. Go work on them.*

Easy homework is not harmless. It changes work behaviors, especially for very capable students. It is yet another reason for us to keep asking what the 'side-effects' of a type of training are. Easy homework was introduced in an attempt to *level the field* in education, and reward students for working, not for innovating. Obviously, since innovating is perceived as a rare ability, there is an expectation that most students will be discouraged and frustrated by a system that constantly places difficult tasks in front of them. The negative aspects of giving students plenty of easy homework go far beyond mere boredom. *Faced with the task of continuously churning responses on easy homework, students will adapt in ways that are detrimental to the goal of becoming good problem solvers.*

Boring, repetitive, uninteresting homework ends up creating a need to do something else and dulls the imagination. Students will incorporate such homework in a stream of multitasking that does not allow for narrow, deep focus on any one task. Moreover, it creates work habits that prevent the development of focus and stamina. It is possible, though I'm not aware of such research, that some students labeled as suffering from ADD and ADHD may be long term victims of hours and hours of easy and predictable tasks.

It appears obvious that we have to *build up the capacity to focus*, not undermine it. The sheer volume of work alone does not improve concentration - on the contrary, the more repetitive tasks we are forced to complete, the more mechanical our responses become. *We cannot improve students' concentration just by assigning a lot of work.*

To improve concentration, the homework itself must be carefully constructed to elicit a creative response. The creative response happens if the task is difficult in any of a number of ways. Either the task is large, such as adding together many numbers, and a clever way of reducing the number of operations is needed, or the task is too complex and the students need to find a clever way of breaking it down into smaller tasks.

# Embracing Difficulty

*You can get help from teachers, but you are going to have to learn a lot by yourself, sitting alone in a room.*

– Theodor Seuss Geisel, a.k.a. Dr. Seuss

After starting problem solving lessons, a father sent me the following lines, where the emotion comes through in every word: "I am certainly feeling the struggle. I am not sure how to ease out this learning. Is there any video based training or anything that can help her connect to this subject matter? The complex problems, I'm feeling at times, make her feel aversion for them. What can I do?" We can accept the fact that it is a difficult task.

His daughter continued classes in which I introduced some rules for interaction that support the acceptance of difficulty, such as:

- It is okay to feel that a problem is difficult.
- All problems in a worksheet should be attempted, regardless of difficulty.
- It is okay to not be able to solve all the problems you attempt.
- All the work you do in your attempts to solve is useful, even if you were unable to finalize a solution.

## DIFFICULTY

- It is okay to not know the answer to a question the instructor asks.
- It is okay to ask a question or acknowledge that you don't understand something about the problem or topic at hand.
- Making an error is an opportunity to learn - share your errors with the instructor or with the other students in the group!
- Having a question is an opportunity to learn - ask questions!
- Even not so great ideas can teach us something about the problem - share all your ideas with the team!

In group settings, I also use some rules that are meant to render the work of the whole group more efficient:

- Problems should first be attempted individually, not as a group (getting an idea is an event that can only happen to one individual at a time; group discussions are often distracting and hinder the generation of ideas.)
- The results of an attempt to solve should be shared within the team regardless whether they represent success or failure, confusion, uncertainty, difficulty, or error.
- Group members should ask questions and the instructor may moderate a discussion in search for an answer (or may answer the question.)
- Group members should share the errors they may have made as this will grow the overall level of experience in the whole class.

Six months later, the daughter could not get to the whiteboard fast enough to show what she had discovered: how any positive integer can be written as a sum of distinct powers of 2. By accepting the difficulty of the problems, she had managed to open for herself a new world of insight which, at her age, felt akin to having magical powers.

## DIFFICULTY

Parents generally think of problem solving as a uniform, undifferentiated activity. They believe that the difficulty of a problem can be set to any level - in other words, they believe that difficulty is a linear function: it can be dialed up or down by the examiner. But difficulty is not a knob that can vary continuously from min to max.

Many parents ask for worksheets ordered by increasing difficulty because they want to have a simple way to monitor the student's progress and a simple way to minimize the time it takes to figure out whether a problem is too difficult or not. In this way they can cap the time spent in unsuccessful attempts, since they perceive unsuccessful attempts to solve as a waste of time. This, however, is not at all true.

Difficulty can be increased in several ways. Of these, increasing the mechanical complexity of the computation is the most likely to support gradually increasing difficulty. For example, an exercise that requires computations with numbers and fractions can be made more and more difficult by adding on more operations, more parentheses, more types of numbers (decimals, repeating decimals, irrational numbers, etc.) But another way of increasing difficulty is to introduce a creative aspect to the problem. Students must make a new logical connection within the knowledge they have. In this case, it is almost impossible to construct a gradual increase in difficulty because the instructor or examiner cannot accurately anticipate how easily or quickly such a connection can be made. Problems are more difficult when the span among the notions that have to be connected is larger, when the solver has to imagine a strategy with more levels, all of them creative. Excepting the most basic applications, the difficulty of a problem is often an elusive characteristic.

In most contests, the difficulty of problems will be structured in three main sections. The first group of problems test for the ability to perform some operations, requiring direct application of textbook concepts. This is followed problems that require a multi-step strategy and the application of a number of different concepts. Some of the problems in this category will require taking some shortcuts or steps that are more specialized, but are nevertheless 'standard.' In the last portion of the contest, problems will require very creative thinking, some ad-hoc development of a new concept, as well as rapid execution. I would like to point out that it is not the same difficulty that increases through these levels, but entirely different types of difficulty. It's not really like the temperature of the oven that we can set higher. The nature of the skills required is different, even as the skills build upon one another.

As soon as the problems require a leap in creativity, parents and students alike find the difficulty somehow 'unfair.' The feeling of unfairness comes, probably, from the fact that they perceive that it is not a linear, 'expected,' increase in the complexity of the simpler, more doable problems, but a leap into territory for which the previous problems have not prepared them.

*In contests, the simpler problems do not prepare one for the harder problems. The harder problems require different, new skills, for which only difficult problems can prepare.*

No amount of simple problems will be sufficient preparation for difficult problems because difficult problems require a leap in the skills and behavior of the solver that easier problems cannot exemplify. The usefulness of simpler problems is limited to building up the ability to *execute* solutions of more difficult problems *after* the solver has come up with an idea - i.e. practicing simple exercises requiring multiplication ensure that the solver will be able to perform the multiplications needed during the execution of a more elaborate solution.

What are the special skills that difficult problems require? First off, an ability to take risks: solvers must imagine an approach which may or may not be successful in the allocated time, they have to try out a few steps at a time and decide (repeatedly) to continue or to change the strategy. Let me remark that there is not a question of confidence. Solvers cannot be confident that a solution will be found, top problem solvers' 'confidence' represents their drive to improvise, build, and assess progress in search for an otherwise uncertain goal. This is the first characteristic of a good problem solver and the point at which beginners turn back: tolerating the uncertainty of the outcome. Going back to our previous discussion about easy homework, we now see concretely how easy homework does nothing to support an increase in the students' risk tolerance.

Secondly, top problem solvers demonstrate creativity when solving problems: they imagine new ways of manipulating data and they are able to re-purpose procedures they have encountered in the past to new, unexpected situations.

Thirdly, the top problem solver has both good time management skills and the ability to work without having an estimate of the time needed to solve. This uncertainty is something the problem solver faces especially when working on difficult problems. Students who solve at a lower level of difficulty typically have a good estimate of how much time it might take to solve a problem. To such students, attempts to solve a difficult problem often seem like 'a waste of time' because it is not clear if and when the activity will end. The lack of such an estimate deters them from the solving activity. However, taking the risk of 'wasting time' is precisely the skill we need to build. This is the main reason why assigning worksheets 'in increasing difficulty order' is not beneficial to the training process.

Lastly, the top problem solver continuously refines the ability to make decisions when solving difficult problems, as well as the ability to make these decisions under stress, by participating in competitions.

There are many behavioral caveats of the 'worksheet in increasing difficulty order.' The most detrimental aspect is that students adapt to the order behaviorally: they solve only the easier problems at the beginning and then find reasons to forgo the rest of the problems. As soon as they identify the first problem that is outside of their current range of abilities, they get a *signal that further efforts might not result in accomplishment* and they stop working. In other words, such a worksheet helps students remain risk averse by telling them clearly where the risky part starts.

When working on a worksheet arranged in random difficulty, students might have to try all the problems, since there is no clear indication of 'where to stop.' This actually helps students attempt new

## DIFFICULTY

problems with an open mind. As a matter of good training, the worksheet in random difficulty order asks students to adopt work habits specific to the top problem solvers: attempt the problem even without knowing from the start if it is doable, invest time in understanding the statement, get used to the uncertainty of success or of how long it will take to finalize a solution.

One of the stumbling blocks in the endeavor of solving difficult problems is the tendency to avoid frustration. If students cannot find an answer immediately, they will ask for help instead of making an honest attempt at understanding the data. A problem that takes a long time to solve can make both students and parents feel frustrated. Spending time attempting a problem again and again, maybe without getting any closer to solving it, can easily be perceived as pointless and unrewarding. This perception hinders students from making progress in problem solving.

In the field, parents ask me "How much time does the homework take on a weekly basis?" They ask this question because they feel a need to allocate and cap the time spent on the activity. However, the attempt to control both the difficulty and the time is detrimental to the training process.

On a small scale, solving difficult problems is a process similar to that of scientific discovery. The genuine scientist is driven by a desire to find the answer to a question, not by a desire to obtain 6 points for that answer. For this, the true scientist does not care whether the problem is difficult, or whether working on it takes longer than estimated. In fact, most major scientific discoveries have not been made on a schedule. True scientists are not daunted by the difficulty, or the apparent impossibility of the task before them. Students should be trained to accept risk and to work in spite of their uncertainty regarding the outcome.

When training, we must focus on the process (have we attempted the problem?) instead of on the outcome (have we found the answer?) without limiting the time we spend in repeated attempts to solve, in researching the topic, or in improving a solution we already have. How many students solve an already solved problem *again*, in an attempt to find a more clever solution? The inability to solve a problem should not be perceived as a failure, but as an *incentive* to research, tinker, and invent.

'Frustration' is a manifestation of discomfort - but the discomfort does not come from the difficulty of the problem, it comes from the cultural expectation that a solution will be found *quickly*. If our cultural premise changes so that we expect difficulty and reward persistence, repeated attempts, and time put in researching, the discomfort will disappear. Of course, we can also make the discomfort disappear by assigning only easy problems.

Parents should ask themselves how they can accept into their modern lifestyles, which are increasingly built around a scarcity of time, an activity that does not have a cap on duration. Parents want to see a math worksheet started at 6 and finished by 6:30 so the student can now move on to the next activity. They do not want to wait until the student 'figures out' a solution, because there is no time limit for this to happen. The fact that ideas do not come on schedule annoys them and they are the first to find this *frustrating*. This is where, culturally, the parent creates and fosters a climate that labels difficulty and the need to innovate as *frustrating*. Instead, the parent should welcome and praise the courage needed to attempt the difficult, the impossible, the uncertain. Only then will students live in a culture that encourages problem solving. Students would never equate difficult with 'bad' if parents would not educate them to shun difficulty.

No parent would concede that it is the culture they build at home that prevents the student from seeking excellence in problem solving. This is because they genuinely believe that by purchasing classes, driving the student to class, hiring tutors, signing up for video lessons, etc. they encourage problem solving. But it is not so, because what they strive to provide as support, they quash by building a defective culture.

What would be a good way to mitigate the sudden difference in skills and behaviors that exists between simpler problems and difficult

# DIFFICULTY

ones? What is the magical bridge that leads from simple to complex in problem solving? In problem solving there is a level at which creativity, not execution, becomes the primary skill. *The correct way to train students is to treat all easy problems as difficult.* The goal of solving easy problems should not be that of finding a correct answer, but that of finding a correct answer in the most clever way possible.

Our work on any specific problem should not end with a correct answer, but with a variety of solutions that all lead to the correct answer. By comparing these solutions we gain insight in the decision making process used to solve more difficult problems: which strategy works well in which context and what the time/execution advantages are of using one method over the other. Every 'easy' problem must be thought of as a part, a component, of a harder problem. From this perspective, we see how important it is to optimize the solutions to simpler problems. Every time we try, to the extent possible, to put forward a clever solution for any problem we attempt, we train towards the goal of becoming able to solve difficult problems.

Top problem solvers do not perceive the difficult problems as easy, but they attempt them nevertheless. They are confident not in their ability to finalize the solution and provide an answer, but in their ability to tinker with the problem and come up with ideas.

So, what do top problem solvers do if they cannot immediately come up with a solution strategy? There are many things they could do: more research in the topics that seem to be related, identifying facts that they need to learn, studying these facts and looking for connections to the problem at hand, analyzing solutions to similar problems that others may have made public, or perhaps *relaxing* and coming back to the problem again after a short while.

These crucial problem solving behaviors can - and should - be used at all times, whether or not the problem is easy to solve. The goal of using a problem set is not to *finish* it but to *study* it, using each problem to expand the set of connections among facts, and to practice fragments of proofs that may be of use as parts of more difficult problems. Seen from this perspective, even the multiplication of two numbers can be interesting. For instance, what we can say about 204 × 197? We can say that the product must be even because at least one of the factors is even, it must end with the digit 8, it will be a 5 digit number, and it will probably be easier to multiply as $(200 + 4) \times (200 - 3)$. First, $200 \times 200 = 40,000$, then we notice that we add 200 four times and we subtract it three times, which leaves just 200 to add to the 40,000. Lastly, we have to subtract 12: $40200 - 12 = 40188$. Because 40,000, 200, and 12 are all multiples of 4, then the result must also be a multiple of 4. This can also be established by noticing that 88 is a multiple of 4.

*DIFFICULTY*

# It's Not in the Syllabus

*The teacher must adopt the role of facilitator not content provider.*

– Lev Vygotsky, Mind in Society

Parents typically have the feeling that, if only they had the 'right' syllabus, it could simply be taught to their students in a straightforward way. First, let me say that yes, there is a list of topics that the student must have a good working knowledge of.[2] Mathematics competitions require all the mathematical topics studied in school up to, but not including calculus. A large portion of the material is normally studied in regular school: pre-algebra, algebra, geometry, and pre-calculus (trigonometry, polynomials, complex numbers, sequences, probability, co-ordinate geometry, conics). In addition to this, competitions ask for a rather advanced knowledge of number theory, advanced counting, graph theory, logic, as well as some special topics that are not covered in school and can form the basis of either independent study, or extracurricular tuition, or both. Geometry is required somewhat ahead of the regular schedule and it is customary for math competitors to study geometry in middle school rather than in high school - this is not unusual in other school systems, such as those of Europe and Russia, where Euclidean geometry is normally studied in grades 7 (2-dimensional) and 8 (3-dimensional).

---

[2]This book offers such a list in the Appendix, for your reference.

## NOT IN THE SYLLABUS

Imagining the syllabus as a list of topics that can simply be covered from start to finish is simplistic. Whenever we study through a list, the desire to get to the end engenders a tendency to study superficially - in fact, the further ahead we get, the more superficial the study. I believe all of us can recall an instance when they decided to study a topic, did a very thorough job on the first few chapters, then spent less and less time on each section that followed, and ended shelving the second volume unopened.

Even more simplistic is the impression that study must always progress to ever more advanced topics. Parents generally resent the idea of going back to previous topics in order to study them in more depth. This is in contradiction with *mindful repetition*, a training method that I defined earlier. I think that a comprehensive math syllabus should be constructed around mindful repetition: some topics should be studied multiple times, each time in more depth and with more connections to other topics.

In many cultures, people think it is not honorable to admit, at the end of a class, that they have not understood and internalized everything. Such a belief prevents the student from asking for clarification or help, thus promoting a certain level of dishonesty in the interaction between instructor and student. Students who have been educated to accept only success interact minimally with the instructor: they listen to the lecture while attempting to look engaged, they never ask questions. If they have not completed the homework it is always due to lack of time, not inability to solve. They resent any interaction that might result in less than complete success: questions from the instructor, requests to come and present an idea, or group discussion of an error they made. These students operate in damage control mode on a constant basis. Students who have been educated to ask questions and learn from errors fare a lot better and make considerably faster progress. Why? Because the problem solving process is not a process of simply knowing the correct answers to questions, it is *a process of continuous improvement on an idea*. When we solve, we come up with some starting ideas: some will be good, some will be mediocre. The mediocre ones are not garbage! They are the first steps to improvement. Problem solvers have to accept that no idea or solution is ever 'the best one.' Whenever we have found a clever, wonderful solution, we might very well find out that someone else has a different one: one that is shorter or more efficient. Parents and educators should educate students to thrive by learning from errors, sharing errors with others (errors are the best teachers!), asking questions, making suggestions - generally, by being active participants rather than passive ones, even at the 'risk' of perceived failure.

A lot of parents encourage and push their students to 'accelerate' through the school curriculum. When seeking extracurricular math activities, parents have a tendency to enroll in ever more advanced classes. Students come and say 'Do we need calculus for this contest? Because we have already learned everything else and we are still not able to do well.' They do not need calculus, they need to come to terms with the fact that they have not learned the material well enough. A culture that looks down on having to re-learn something impedes possible progress. The expectation that learning must always be 100% indicates to students that the only direction is forwards, to ever more advanced topics. The idea of 'going back' feels like a stigma, even though with each pass through the material we acquire a more detailed understanding of it. It is absolutely normal for the learning process that students do not - and cannot - retain 100% of all details at the first encounter with the material. The reason for this is that it is very difficult to perceive the relative importance of the details and the need to internalize them. Students normally focus on retaining whatever seems more important and, in judging that importance, they are hardly experts. Furthermore, the importance of each detail is not immutable: it varies in time according to the needs of the sciences that use mathematical models. Some mathematicians have had remarkable careers by discovering how to apply to modern technology facts that had waited for centuries in the dusty recesses of mathematical literature.

The fact that parents keep pushing for keywords such as 'advanced', or 'forward' is a cultural aspect of learning mathematics that results in a lot of poor decisions in managing math education. It is important to study the same material more than once, perhaps from different angles or at different levels of detail.

## NOT IN THE SYLLABUS

Because problems span diverse notions across the entire syllabus, a problem solving activity cannot be entirely systematic: we cannot make it progress along a list of concepts. Problem solving, as well as mathematics as a field of knowledge, is a web of concepts that cannot be represented as a single thread. We often find that, in order to understand a topic better, we have to expand our knowledge of other related topics as well. Most steps forward are accompanied by several steps laterally, or even back. For example, to solve a geometry problem, we may need to expand our knowledge of quadratic equations.

Only the simplest problems are based on a single topic. More advanced problems combine notions from different areas of mathematics and require solvers to discover connections that they may never have noticed before. Good competition problems are meant to give opportunities for creative solving and are, therefore, quite unique. There is simply no way someone can *learn how to do them all* but, rather, one has to learn techniques that are powerful and general enough to inspire a creative solution.

Another barrier to the 'topic-by-topic' approach is the difficulty of the problems. Let us not forget that the problem solving skills required to solve a specific problem depend on its difficulty! So we may imagine learning geometry based on an array of simple applications but, as we progress to higher level contests, we will find that a second or even a third pass through theorems is necessary in order to improve the observation of the relevant details.

At this point we might conclude that it is more efficient to train in reverse: start with the problems and learn theory as needed. Study in order to understand the statement. For example, if the statement mentions angle bisectors, go and learn or review everything on angle bisectors, and so on. Then take a few steps through the solution and, if you find yourself 'stuck', learn or review whatever topic you are stuck on. In practice, this manner of covering the syllabus does not completely work. Its attraction consists in the fact that it promises to build problem solving skills while allowing the student to learn the much needed knowledge in a manner that exposes the creative connections between notions. Its caveat is that in many cases, especially in more difficult problems, it is not at all clear from the statement which concepts are needed in the solution. Students might very well encounter a new definition that the problems they solved previously did not prepare them for. What if the problem statement does not *mention* angle bisectors, yet they are somehow hidden in the wording? For instance, if the statement tells us that we are dealing with a line that is equidistant from two sides of a triangle. Only the student who is very fluent with theory will identify the description of an angle bisector.

It is not a simple matter to cover the entire syllabus *efficiently* but one must keep in mind that, to advance as a problem solver, it is not sufficient to move to ever more advanced levels of theory. Simultaneously, one has to advance the level of difficulty at which one is able to solve as well as the ability to navigate connections that bind together the different concepts that are used by an interesting problem.

*NOT IN THE SYLLABUS*

# Speedy Math

*It's not that I'm so smart, it's just that I stay with problems longer.*

– Albert Einstein

I was born and raised in a world without computers. Back then, as the first steps towards automation and robotics were partly real, partly science fiction, and Stanislaw Lem imagined fairytales and legends of artificial beings, there was no question that robots and computers should be designed to be as humanoid as possible. We modeled everything on ourselves, and, perhaps too proud of our own intelligence while on the brink of unleashing the power of digital technology, we dreamed of androids and AI that could feel emotions. Forty years later, teaching gifted students in the San Francisco Bay Area, I'm noticing a paradoxical trend: fascinated by technology, immersed in a world full of powerful devices, we now dream of becoming machines. Our children are treated as imperfect units that, trained appropriately, must become fast and accurate executants.

The main difference between how a calculator operates and how a human operates is, of course, intelligence. And, while there is much talk about artificial intelligence, many people do not care to find out how artificial intelligence *differs* from natural intelligence. Though progress in developing artificial intelligence has been very rapid, humans still continue to have the unique ability to *create* rather than just *execute*.

In meetings, I often point out to parents the difference between finding a correct answer by plugging values back into the problem (the famous 'trial and error' strategy) and doing so by applying observation, logic, and creativity to the given data. Any machine can be programmed to solve problems by 'trial and error.' Machines perform arithmetic operations much faster than humans and can try out values until their program finds that the value is sufficiently close to the solution. Nevertheless, machines cannot yet come up with a better way to solve a problem. Students who solve using machine-like techniques are at a disadvantage *in the long run*. To make matters worse, both they and their parents conclude that they are on the right track to becoming good mathematicians because of the number of the correct answers given. What is wrong with becoming more machine-like? What is wrong with attending math programs that constantly upgrade our 'math processing chip'?

This is what is wrong. *Students will 'max out'* - that is, their capacity to rapidly test numeric answers against the requirements of the problem is limited. There comes a point when the upgrades fail to help and the size of the problem clearly requires a more clever approach. Students will not be prepared for the occasional never-seen-before task and will feel less and less confident as their scores systematically fail to rise. Worse even, their success in finding correct answers does not represent an incentive to become creative. Such students start to manifest little or no interest in 'clever solutions' and perceive them as unneccessarily difficult ways of 'finding the right answer.' Sadly, it is only too often that I find students who perform extremely well in middle school math but who do not work based on proof and logical reasoning. Generally, their ability to solve problems evaporates as the complexity of the problem increases.

I remember a student who, when competing in the American Invitational Mathematics Examination (AIME) wrote up six pages of numbers to find the answer to a problem about sequences. I think of this as an extreme case of brute-forcing. However, certain aspects of his approach paint a more complex picture than just a black and white argument. He made the decision that the problem exceeded his capacity to find a proof-based approach to solving. He decided to list all the terms of the sequence up to the needed term. As a result, he used ten or, perhaps, fifteen minutes to cover six pages with numbers with flawless accuracy. A single error early on would have affected all the subsequent values. After this ordeal, he did find the correct answer. It is virtually impossible to say how the correct answer was found just from the answer alone. Had he had less time, or had the solution required writing sixteen pages of numbers instead of six, the story would have had a different ending. A proof-based solution can be reused, expanded, or modified.

Generally, any valuable math education program should focus on *creativity* rather than mechanical ability to solve. Unfortunately, many programs that claim to focus on creativity result in a state of chaos. This is primarily a consequence of believing that, by giving humans every opportunity to create, they will do so. This is not necessarily true. Often, a stress factor is required to jumpstart the creative process.

Of a number of students asked to rapidly compute $202 \times 202$, a few students will do $202 \times (200+2) = 40400+404 = 40804$ rather than perform multi-digit multiplication. On the other hand, if students are asked 'how would you compute $202 \times 202$ in a clever way?' and given plenty of time to come up with an answer, students will just sit and wait to be told how. Not being subjected to any burden, they will not feel an urge to simplify the task.

These aspects of student behavior make math education a highly non-linear task. We cannot simply decide that we will avoid 'dumb, repetitive tasks' on principle. At the same time, we cannot simply subject students to 'dumb, repetitive tasks' as a means of provoking a creative response in some of them. At least in my experience, students adapt in different ways to this hardship. Some of them train themselves to calculate extremely fast without finding any clever, creative shortcuts.

# SPEEDY MATH

The drive to ask students to perform operations mentally at high speed, followed by speed contests such as 'countdown[3] rounds,' is intended to spur a need for creative response. Questions such as: 'Calculate $2015 \times 2016 - 2016 \times 2014$' are meant to force some students to factor out 2016 and realize that $2016 \times (2015 - 2014)$ is simply equal to 2016. However, the students who answered correctly, may have used any of several processes: some have simply multiplied and subtracted mentally (adapted by becoming more calculator-like), some have suddenly 'discovered' the factoring shortcut (adapted by becoming creative), others yet have been trained from before to factor in such situations.

Of course, in life, we are interested only in correct answers and almost not at all the process though which they were obtained. However, education differs from actual professional practice: it is not a small scale model thereof. Often, the process of education is non-linear, and we have to take a weaving zig-zag path to a goal rather than a straight one. If it is desirable for a professional to be creative, then we have to reward the cleverness of the solution more than the correct answer itself. It is my belief that, if a person is smart, creative, and humble, they will learn when they have made an execution error and they will be able to fix it.

Leaving rapid calculation aside, how does a student learn how to solve problems *faster*? Definitely not by putting special effort into this aspect. Parents often think that, by learning a lot of rules and practicing exam problems regularly, the student will be able to speed up their process of solving. This is not quite true. Let us remember that the more difficult problems require some innovative step, some creative extension of previously learned facts. To solve any of these hard problems the student must come up with an original idea. How do you make *that* process faster? To 'speed up' the pace of solving we have to focus primarily on the level of detail of our learning: the less mechanically we work and the more logically we try to internalize each concept (by proving it from scratch), the better we get at *generating*

---

[3]'Countdown' is the name of a speed-based round of questions in contests such as MATHCOUNTS and Math Leagues.

*crucial ideas.* If we, instead of *learning* a fact, *re-invent* it, we build a strong mental library of relations among mathematical facts. In this library, we store *how* everything is proven, not *what* everything says, thus perfecting our skills for proving new facts (the necessary condition for excelling in math competitions and problem solving in general.)

Let us use a simple image: we do not improve our speed at fishing by learning how to throw a fishing line into the water faster, but by weaving a fishing net. The student who learns 20 rules for divisibility with difficult divisors (such as $7, 13, 19$) and practices to apply them quickly will, in the long run, be surpassed in speed of execution by the student who creatively makes use of number theoretical approaches that are suitable for the number at hand.

The education process cannot be all fun and comfort, some difficulty and frustration must be engineered into it in order to channel energy towards being creative. The requirement for speed in computation is one such difficulty. Used correctly, it will help the student create shortcuts. Used incorrectly, it might just turn the student into a machine. The same hammer that is used to build a house can also be used to demolish it.

# Man, Machine, or Both?

*The real danger is not that computers will begin to think like men, but that men will begin to think like computers.*

– Sydney J. Harris

My most vivid memory from the Moscow of the eighties is of the typical cashier. Armed with a wooden abacus, a cashier could ring in a cart of goods faster than a modern point of sale. Notwithstanding her rapidity on the abacus, the fact that she had memorized all the product prices turned checking out of the store into quite a show. I was traveling with a group of physicists and we often talked about the almost magical abilities of the Russian cashiers.

About the same time, TI scientific calculators were starting to become a standard tool for engineers. However, no one really thought that becoming a proficient *user* of the scientific calculator was the same as becoming a scientist. People thought there is more to being a scientist than knowing how to use a calculator, but started to be unsure whether it was really necessary for a scientist to know exactly how to perform basic operations by hand.

Nowadays, everyone with a personal computer can purchase a license for symbolic computation software to have access to a machine that performs algebraic manipulation of expressions, graphs functions, or performs indefinite integration, to name only a few possibilities. To be a proficient *user* of these tools, one only has to identify the operations that are needed.

But in all these examples - from wooden abacus to symbolic computation applications - there is a common definition of the user. The user has to know how to *enter the data* and retrieve the result from the device. Data entry is the step that requires the most specialization. The user does not have to know *how* the machine operates. This is fine, unless we somehow make some sort of a crazy leap of faith and assume that, because one is an expert user of a machine, one is also an expert in how the machine does what it does. It is not sufficient to be able to enter '14 + 97' in order to show that we understand how addition works. This is something everyone can agree with, but some people will say that it is no longer *important* to understand how addition works. They may be right or they may be wrong. But, we can definitely say that, we have no way of knowing what might have happened if addition would have been excluded from the math syllabus when the mechanical calculator was invented.

Many mathematics competitions do not allow the use of calculators for an excellent, albeit often misunderstood, reason. This reason is that *mathematics, as a whole body of science, is a giant set of tools for simplifying computations.* A well designed problem will require students to demonstrate creativity in simplifying computations because this is one of the main jobs of mathematicians: multiplication was invented to simplify addition, integration was invented to simplify addition, geometry was invented to simplify measurement, etc. A well designed problem requires students to creatively simplify expressions (numeric, algebraic, trigonometric, etc.) or to find shortcuts that bypass computation altogether (e.g. Viète's relations vs. quadratic formula, etc.) By having access to a calculator, the student may be able to brute-force the problem, avoiding the need for a clever strategy. This observation brings us back to our main thesis: *problem solving competitions are not about the correct answer, but about demonstrating a creative strategy.* The correct answer is just a metric, a relatively poor substitute of the hand-graded full solution where each student is given feedback about which decisions/ideas/computations may be improved.

Many problems at a higher level of difficulty will require execution that involves simplifying computations. The simplification of expressions, either numeric or algebraic, is the most direct type of exercise that requires honing the skill of always keeping the numbers small. Mastering a variety of techniques for simplification often provides *inspiration* for designing solutions. This is because the task of simplifying numeric and algebraic expressions is not a useless puzzle, it is an exercise that builds and strengthens the ability to solve problems in general. It requires keen observation of details, ability to predict a few steps ahead in the computation, ability to find an optimal order of application for a set of rules, accuracy, and the ability to make decisions, in addition to the main task which is, obviously, computational. Solving computational puzzles without a calculator is not a task intended to train not only the student's computational skills, but also the *other* skills I mentioned.

If calculators are not allowed during the competition, should they be used when training? My answer to this question is definitely yes, in specific ways. Some examples of meaningful use of calculation tools are:

- *After* acquiring proficiency in prime factorization, use a calculator to explore numbers that have large prime divisors. This experiment by no means replaces the need for the student to find ways to optimize the search for a possible divisor using clues such as the last digit of the number.
- *After* studying a type of function (absolute value, quadratic, higher order polynomial, exponential, logarithmic, trigonometric, etc.), use a graphing calculator to explore variations in the shape of the function.
- To *verify* the result of a pen and paper computation.
- To *verify* the result of graphing a function on paper.
- To *verify* the result of solving a set of inequalities on paper.

Therefore, I recommend using calculators and advanced computation software for two purposes: *experimentation* and *verification*. I would like to note that using advanced tools in the study of mathematics *adds* to the time spent studying and does not, as is commonly assumed, shorten this time, or replace the need to train computational skills. Electronic tools open new horizons of exploring complex examples and are invaluable tools for advancing the student's insight into the subject matter as long as they are an *addition* to the training process.

# Keys to Confidence

*Confidence is ignorance. If you're feeling cocky, it's because there's something you don't know.*

– Eoin Colfer, Artemis Fowl

After her first grader, Ken, attended classes for a short time, Janet sent this message: 'The problem is he has never been in a competition and he is too young to understand that it is ok if others go ahead of him and answer. He had tears in his eyes last week when I believe he got several answers wrong. I don't want him to lose confidence. He can get back to lessons anytime if his confidence is preserved.'

Ken had been immersed in a group of kids of the same age and math ability. In any such group activity it is inherent that some will find a solution faster than others. Why it is that children do not start crying if they swim slower than others, if they don't always win at chess, or if they don't make the next level in a video game? How come they can play video games for apparently limitless periods of time and never burst out in tears? How come they don't panic before the next chess class? Is chess less difficult than math problem solving? Why does arriving fifth at the other end of the pool lane not trigger the immediate reaction to drop out of swimming as an activity?

It is a matter of expectations. If parents expect their child to perform better than anyone else in the class, then, of course, every parent but one will see that expectation fall short. Families tend to make a big deal of a child's early intellectual fitness, fawning over all the proofs of cleverness and, consequently, burdening the child with an obligation to top expectations of intelligence. A parent would never impress upon a child the expectation that they will win at chess. Because chess is a game with two opponents, it is more obvious to the parent that the child may be on the losing side. However, in a math or science class for gifted students, each parent readily assumes that their child must rank at the very top.

A group problem solving activity cannot be totally non-competitive. Some students will find a solution faster than others. Some students will find better solutions than others. Some students will be more articulate when they explain their solution than others, etc. Many activities are inherently competitive and a group lesson for problem solving is one such activity. Parents should educate their children about the expectations of such a class: that they may not be the best in the class, they may encounter difficulty, and they will need to put in persistent and even hard work to perform at a high level. It is not the comparison to other students that distresses a participant, it is the initial expectation that this comparison will be instantly favorable.

It is not correct, nor helpful, to label mismanaged expectations as *confidence*. What do parents mean by this? Do they mean students come to class confident that they will amaze their instructors and the other students? The belief that they possess a super-high level of intelligence, or that their math level is 'off the charts' is not healthy. Parents should educate children to expect that other students may be faster, more clever, or more knowledgeable than they are. This in no way diminishes the value of their work and passion for a field of knowledge.

What does it mean to be *confident* in math? To expect that one will be able to solve the problem? Surely, not even Einstein believed this about himself. It is misguided to believe that great mathematicians and scientists are, as a rule, confident in their ability to find a rapid, elegant solution to any problem. In fact, the more mathematically educated we are, the harder the problems we tackle are, and, consequently, the less confident we are that we will find a solution. Einstein spent a lot of his life attempting problems he died without being able to solve. He was not confident that he would solve them. The higher we go, the more adventurous the attempt. When Poincaré stated his famous conjecture, did he start crying because he wasn't able to find a solution? Did he quit the field of mathematics because he was afraid that someone else might prove it ahead of him? No. Poincaré's conjecture remained unproven throughout his life and, at the end of almost a hundred years of mathematical progress, its proof was finalized by Grigori Perelman, one of the great mathematicians of our time.

*Confidence* is not a sentiment that pervades the population of authentic mathematicians. Moreover, any good mathematician will tell you that being *humble* is a more appropriate feeling when attempting a problem. True intellectuals must be able to acknowledge errors, fix them, and discuss them openly. *Bragging rights are not the goal of problem solving!* Good scores at a math competition are not a reason for feeling and behaving like a celebrity! Living with the fact that an intellectual challenge may not be surmountable is part of a scientist's education.

For a scientist, *confidence represents the courage to attempt problems*, not the actual ability to solve them. Because it is inherently uncertain whether a problem can be solved upon a specific attempt - examples of problems that have taken decades of work done by several generations of mathematicians abound - a scientist must be trained *to try*. Placing the focus on success at every attempt is detrimental, because it effectively deters students from trying. In the long run, the more parents focus on success, the more paralyzed the student's decision making ability becomes. True *confidence* is the ability to keep trying - in spite of perceived lack of preparation (we can learn, we can better ourselves), in spite of the perceived difficulty of the problem (we may not be able to solve it completely, but we may be able to go a bit further towards a solution and thus *help someone else* solve it), in spite of the perceived lack of time (another generation of scientists may finalize it), and in spite of a perceived lack of resources.

Janet's son lost the *confidence that he was among the best*. By participating in class, he started to build the *confidence that he is ready to attempt a solution*. Like cholesterol, there is good and bad confidence. Losing the confidence that one is 'good at math' is the first step on the path to becoming better at math. It is necessary but it does not have to be painful. It is made painful by the culture of success and excellence that the student has been brought up in, not by immersion in a group with a competitive dynamic. The wise student enrolls in a class to improve, not to show off.

I often hear from parents about their motivation to sign up for a math enrichment class. On occasion, the reason is similar to this: 'His results on tests are "off the charts." I thought I should sign him up for some advanced classes.' This conclusion is based on school test scores and has a few caveats. First of all, tests are increasingly easy lately, and entirely based on the simplest application of recently learned facts. What else can make a parent happier than an official score that shows their child among the top 2% of all students? Though it is difficult to imagine anyone scoring badly on these tests, experience shows that, in the general population of students, the most frequent behavior among disinterested students is to *check answers at random*. School districts then go and perform automated *item analysis* on the test results and come up with various ways for improvement. Of course, since the answers were picked at random, the analysis of the items will show little pertinent information. Finally, the district will conclude that, no matter which items are changed, the statistics remain the same and the only way to improve the scores is to *lower the overall difficulty of the test*. Over the years, these blind policies have managed to bring tests to ever lower standards. It is not difficult to score 'off the charts' on a test if many students check answers at random.

Though the decision to sign the student up for some enrichment classes may be correct, the expectation that the student will continue to be in the top 2% of the class is going to result in disappointment. It is very likely that the other students who have signed up are in a similar situation. The majority of the class will consist of students of above average intelligence who are more motivated to study mathematics than the general population.

It is also noteworthy that there is a factor of luck in every success. Every time we get lucky, we accumulate more of the bad confidence: the confidence that we are special, that we can do well by just showing up when others work a lot only to fail. Our society as a whole celebrates and rewards lucky people. As a form of entertainment, this is as good as anything else. However, within the realm of education, of culture that affects the dynamic of a family, it is detrimental to praise and reward luck. What should we then reward? The ability to work, attempt to solve a problem, fail, yet go back to work and make some progress. *Progress in one's work* is the highest expectation we can have from a scientist. *Success in one's work* is the highest expectation we can have from fate. It is not success, but progress, that builds the 'good' confidence in one's skills.

# Creativity in Problem Solving

*One must still have chaos in oneself to be able to give birth to a dancing star.*

– F. Nietzsche

When I taught my first class in California I noticed that students often wrote 'IDK' beside problem statements. I did not hesitate to ask them what they meant by this. Peals of laughter later, they revealed to me the meaning of the acronym: 'I don't know.' This puzzled me:

'You don't know? You expect to *know* how to solve a problem? If you knew how to solve it then it wouldn't be a problem, I think. You have to *figure it out!*'

It turned out nobody had ever told them that they were expected to *figure it out*. They believed that they had to *know* what to do and that the responsibility for making that happen was mine. Was I in possession of the algorithm and was my task to program their heads? Yes, if we express more crudely what a school district official told teachers at a meeting I had the chance to attend: 'With the help of the Department of Education from X. University, we will produce a step-by-step lesson plan for each math lesson accurate enough to enable any person, even without qualifications, to teach the lesson. In this way it will be very simple to substitute.'

Generally, it is not the ability to execute, but the creative aspect of the work that constitutes the main difficulty of teaching problem solving. Solvers will perceive the problem as difficult only when they 'get stuck' and 'do not know what to do.' This type of feedback reflects a cultural aspect of math education: the student expects to *know what to do* when, in fact, the expectation is that the student should *figure out what to do*.

As a keyword, *figuring out* has somehow been relegated to the dustier sections of the vocabulary, because it cannot be measured objectively, except by waiting to see if someone actually produces an original solution. This is an event that humans admire and remember in history books but do not, generally, promote culturally. The unpredictability of the creative thought is a subtle poison that laces the ambrosia of genius: a genius may produce something or nothing. What society generally wants from the individual is to earn a monthly paycheck and fit well in the corporate structure. Instructors and exams want to assess quickly, most often mechanically, if a student is proficient. Bosses and investors want deadlines by which they can 'see results.' Society does not have any real desire to foster creativity while also, paradoxically, having an immense appetite for it. Large corporations often purchase innovation from start-up companies, who can 'afford' to take the risk. This is not necessarily bad, but it reveals the chasm between society as a whole, which is very risk averse, and its creative entities, which inherently carry risk.

The impression that one has to *know* the answer must be the first to go. A problem is a problem because, and only because, we do not know how to solve it. If we did know how to solve it, then it would not be a problem, it would be an exercise. Since we cannot know from the start how a problem should be solved, what we have to do is *figure it out*. The less obvious it is how to figure out the solution, the more challenging and interesting the problem. The distinction between a problem, which needs more thinking and researching, and an exercise, which is an immediate application, is not absolute; it is relative to the student's training and ability to solve. The same statement may represent both a problem for one student and an exercise for another student who is a more advanced problem solver.

Creativity in mathematics is often thought of as a high level process, in which the creator manipulates the so-called *big picture*. This may or may not be true, but even if it were true, it is definitely out of the question that a person could manipulate the big picture without having a very, very thorough grasp of the concepts and details that underlie it. Parents who think it is possible to function at a level at which only talent, but not difficult, 'boring' work, is needed, are considerably off course.

It appears therefore, that *difficulty* and, perhaps, *frustration* are feelings that occur naturally before we summon up our energy and inventivity. To foster creativity we must accept difficulty as the cornerstone of meaningful assignments. We have to also accept the *frustration-innovation* duality. When human intelligence feels frustrated by an apparent impossibility, it innovates! Unlike machine intelligence, *human intelligence works together with an array of emotions* to generate creative responses. It is the co-operation of emotions and intelligence that makes us human and, for the time being, different from intelligent machines. A creative response does not occur in a predictable manner but the emotional factors that may elicit it can be identified and introduced in our culture of education, instead of removed from it. Curiosity, ambition, or frustration are feelings that may motivate a creative response. Frustration with not being able to accomplish some goal is probably the most fundamental *emotional driver of creativity*.

Problem solving competitions offer parents a way to verify, from early on, how important it is to develop the ability to explore the details, rather than ignore them. *Creativity happens only when there is a large accumulation of knowledge and an emotional drive to connect parts of the knowledge in new, original ways.*

Let us assume we are now at the point at which we have replaced *I don't know* with *I haven't yet figured it out,* and we are ready to talk about what it takes to actually find a solution. If a good deal of creativity is needed, how does the student develop such an ability?

To become creative one has to not only be diligent but also curious. Creative students are intrigued by problems they cannot solve immediately. After devoting complete attention to the problem for some time, they memorize it and keep thinking about it while engaging in less demanding activities. At times, an idea comes to them and they go back to solving with renewed interest. Unfortunately, thinking is an untrusted activity at the end of parents, teachers, and work supervisors who like to *measure* work and progress. After all, thinking is hard and, when observed from the outside, it is practically indistinguishable from idling. Spending time to figure out a problem without producing an answer can be perceived as frustrating by both students and parents. However, such 'stalling,' can mean either that the student has given up on the problem and is now waiting for help or that the student is making repeated attempts to solve the problem. The latter behavior is part of the process of learning how to solve! Students are hardly going to become more creative if they are constantly shown how somebody else has been creative. 'Getting stuck' is good - it is an incentive to make that leap from what we know to what we create from scratch.

A student cannot have a genuinely creative experience in the half hour left free between ballet class and orchestra rehearsal. Thinking, especially deep thinking, requires time to execute, just like any other activity. The *greed for structured educational activities* goes against what most parents aim for: to provide high quality, structured activities that fill up the after-school time of the student. School, as well, fills even more time with homework, projects, and events. As a result, a student in the San Francisco Bay Area has hardly any free time for self-directed activities. As far as society is concerned, this keeps students off the streets and out of trouble. At home, highly educated parents realize with concern that school does little to actually teach skills and try to compensate by using tutoring centers, college sponsored camps, and internships. Together, school and parents leave students no time to dream, think up stuff, figure out things, or read books that are not on the school-recommended reading list. A former student of mine, for instance, taught himself Russian in order to be able to have direct access to Russian mathematical publications. Giving the dedicated student just this kind of free time can have amazing results.

I routinely interact with students whose typical Sunday starts at a tutoring or SAT prep center at 9 AM and ends there at 8 PM. Parents come and drop off a bag of take out around lunch time while the student attends classes back-to-back. In other cases, students spend their weekends being driven from one activity to another in an endless string of personal improvement: from dance, to math, to piano, to skating, to chess, to debate. Often, activities encroach upon one another: the piano competition conflicts with the dance class, the dance recital conflicts with the math lesson, and the chess tournament conflicts with debate practice.

For many years I have heard parents excuse absences. Surprisingly, I have also had parents attend class instead of their children - they took notes and hoped to explain the lecture at home. Students attend math classes with Nutcracker ballet costumes on, with stage makeup on, or with their skis by their side, ready to run to their vehicle and be whisked to their next activity. What is it that these privileged students do not have? They do not have the time *to think their own thoughts*. They are supposed to learn continuously and never have any time to take stock of what it is they are learning. It is no wonder that some of them doze off during classes, use their phones and tablets to play and use social media, or simply use class time to chat with their neighbor.

Another detrimental side of this type of parenting is a lack of tolerance for any non-conformity. The student is not allowed to go to school with unfinished homework even if this means getting fewer hours of sleep. An environment which deprives the student of sleep on a regular basis is not supportive of long term progress in problem solving because sleep is a crucial factor in a lot of the brain activities that support problem solving: retention of information has been proven to happen during REM sleep. I hear too many parents explaining how 'She will sleep tomorrow. She has to finish this now, but she will sleep tomorrow.' Students flesh out the picture: 'I haven't done so well. I had only five hours of sleep last night, and the day before only four hours. But I will get some sleep on the plane to the piano competition.'

Students need to sleep and benefit from free time. I agree that, for some students, free time may not be well spent. However, I do not agree with the extreme view that free time should be eliminated because it is uncontrollable.

To discover the roots of creativity we have to travel back in time to the ancient Greeks. After all, they invented number theory and wrote almost all the geometry one routinely learns in school. What is atypical about ancient Greek culture is its unique rate of creative output: a small number of people created an enormous amount of important knowledge that we still study today. What made them so creative? They had plenty of free time, that is clear, but how come they created math, sciences, literature, art, and philosophy with it while other people looked on in awe (including the Romans) without equaling them? The answer is in their *culture*. They built a culture in which creativity, figuring out, and building from scratch, were highly prized and respected activities. Furthermore, their culture placed less value on material possessions than other societies did. The ancient Greeks liked to be rich, but did not earn respect in society solely by being rich.

A culture that supports creativity values *the quest* more than the *success* of the enterprise. This may sound quixotic, but it is also a key to success or, in any case, to happiness in seeking it. Concretely, as far as problem solving is concerned, students should be encouraged to spend time pondering over problems, rather than immersed in a 'correct answer equals good, incorrect answer equals bad' culture.

Correct answers are not the only goal of problem solving. To this end, incorrect answers are opportunities to learn, fix, and create. The goal of problem solving activities is to find a well-reasoned, logical, and elegant way to the answer. This proof-based strategy is what we must share, enjoy, praise, and respect, not the cold numbers of quiz scores and test results.

# CREATIVITY

# Let's Get There Faster!

*He knows so little and knows it so fluently.*

– Ellen Glasgow

Sunitha feels overwhelmed by the number of educational options available for her son, Arun, a very gifted 8$^{th}$ grader. She can enroll him in an 'accelerated algebra II' class, or sign-up for advanced problem solving classes, while Arun himself would prefer to have more free time to put towards studying mathematics on his own. Sunitha feels that letting Arun self-direct his study will result in less overall progress, slower learning, and loss of control. She wishes she would know more about the long term effects of each of the possible decisions, but she has no crystal ball, while the instructors and parties involved offer only biased information, meant to promote enrollment and sales.

Without going deep into the current landscape of mathematical education in schools, I would like to discuss some of the more controversial aspects of it. For instance, 2016 has been the first academic year when students no longer learn how to multiply multi-digit numbers without a calculator. Long division was axed the year before. Technical drawing for geometry is no longer in living memory. If these skills have been removed from the syllabus because they are obsolete, which newer, more useful skills have they been replaced by? None, really. A look at any thick, shiny textbook reveals watered down material interspersed with very simple exercises that are completely devoid of original thinking. Gifted students have no difficulty at all completing chapter after chapter and obtaining an A. Does this mean they are ready for the next level of math? Does this mean they can solve problems based on the acquired knowledge? Does this mean they can provide creative responses, creative solutions?

Unofficially, the way schools expect mathematics to be taught is generally not based on pre-requisites. Even if, officially, it is acknowledged that the student must complete Algebra I before taking Geometry, principals and teachers try to avoid the need for previously acquired concepts or skills. Therefore, the reliance on pre-requisite learning is reduced to a minimum. As a result, not only are teachers no longer responsible for delivering results that enable students to understand the next level of math, but students are relieved as well from the need to connect concepts, to understand, use, and control the vast network of knowledge that is mathematics. The student can now learn trigonometry without being sure what a circle is (indeed it is amazing how few students, as well as teachers, are actually capable of giving a *definition* of the circle), or learn about domain and range without knowing what a set is. And while this is actually fascinating in terms of a human's capacity to learn out of context, it is very detrimental for problem solving because *math is about learning in context*. Ability in math is the ability to *create connections* not the ability to function without them!

Lilian, a candidate for tenure as a mathematics teacher, once told me: 'My supervisor told me to teach mathematical analysis as if students had never learned any math before. She said one could never assume they had been taught any specific concept in a way they might recall and that I should design my classes using mechanical steps to be followed in response to usual questions.' Regardless of the lofty mission statements of the school districts and departments of education, the ugly reality in the field is that the continued standardization of education - both in content and quality - cannot but be the enemy of creative response. In fact, in the field, problem solving is taught as a collection of recipes to be followed without modification. However, problem solving is, undoubtedly, one of the most creative human ac-

## ACCELERATION

tivities. It is detrimental to pretend it is a routine that can be learned by repetition.

An accelerated curriculum is only a way of saying 'this syllabus has now become so watered down that we can actually wrap it up faster.' In mathematics there is *a lot* that can be researched, taught, and done about and around any given topic. The wealth of connections, variations, alternate solutions, generalizations, makes any mathematical topic difficult to wrap up. A student who is really advanced in the study of mathematics will not want to learn it faster - and, inherently, more superficially - but slower and in more detail. The *acceleration* that we are experiencing in schools is not a method for learning *more per unit of time*, as the name suggests. It is a method for learning less, overall, because each topic takes less time to 'finish' as it is taught in less detail.

Due to accelerated study I regularly see students in grade 8 who have 'completed' pre-calculus with straight A's. However, the vast majority of them do not have a thorough enough understanding of many of the concepts involved. Of course, 'thorough' seems to be a relative notion and this enables school districts and even colleges to turn 'thorough' into 'superficial' with the magic wand of demagogy. But 'thorough' is not a relative notion. 'Thorough' means whatever is needed to *get the job done well, fast, and without errors.* If the job is that of solving problems, then the results of an accelerated curriculum fall far short from 'thorough'. Accelerated students with straight A's come to class to find out with surprise that the base of a logarithm cannot be equal to 1, or that graphs of polynomials never exhibit asymptotic, etc.

Accelerated study gives parents the false impression that their child is ahead in the study of mathematics. Instead, a thin layer of ice has formed a temporary bridge over the unexplored depths of the lake. It is very difficult for a parent to make the decision to 'go back' and study thoroughly material that seems to be a repetition of previous study. A credit obtained from a school or other accredited institution carries a lot of weight and parents would like to believe that it represents a milestone, that with the knowledge acquired the student is actually able to solve problems, as well as rely on a foundation of skills that will last a lifetime. Parents are, naturally, confused - if we have to 'go back' how far back and what kind of resources are needed to turn the bridge made of thin ice into one made of concrete, or, at least, of wood? And I think this is, basically, the turning point, the point at which we need a *real* assessment of the *real* skills a student can demonstrate - this is where mathematics competitions come into play. Their role and responsibility are enormous - they are the only honest skill test available. As the only true measures of ability, they have to provide relevant assessments and selection criteria that help students and parents understand how close or far they are from mastering the topic.

Accelerated study is not a means to prepare for mathematics competitions, at least not in the form that we see it today, as an even more sparse version of an already diluted content. How can we tell if a study is successful? By measuring the *level of creativity* the student can display. Of course, posing questions that test for a creative response is extremely difficult and the task is made even more difficult by Google, the information marketplace where students find answers to problems. The examiner must produce very good quality questions that are *new* all the time so that searches on Google will, hopefully, be unsuccessful. Obviously, this does not happen and cannot happen in the current educational infrastructure. For the most part, the quality of the entire study hinges on the quality of the questions asked. As a trend, the quality of the questions is decreasing, if anything. The only good quality questions out there are competition questions.

Is acceleration useful? Yes, of course. By representing a simplified, more superficial syllabus, acceleration is useful for students who do not plan to have careers requiring mathematical strength or for students who wish to obtain their math credits easier in order to put more time into studies other than math. Students who are good mathematicians, on the other hand, should opt for *more* time spent learning each topic in depth, not less. How much time? Again, creative response is the benchmark - as long as it takes for the student to become able to make new, original connections between the concepts that were studied.

# ACCELERATION

I had a very talented student a few years back, who was invited for the AIME both in $8^{th}$ grade and in $9^{th}$ grade. In his $10^{th}$ grade, his mother insisted that he take lessons in order to have a chance to participate in the USAJMO. After we started lessons, he realized that his AP Calculus BC class was extremely difficult. By relying on a foundation of 'accelerated' credit classes taken in parallel with school or during the summer semester, he became eligible for this rather demanding calculus class two years ahead of time. Little by little, his ability to find time for solving problems at a highly competitive level started to erode and our lessons went south, dealing mostly with review of the topics he had not had sufficient time to internalize. To top it all off, piano auditions and competitions were coming up. Soon enough, we were going in circles: each week we were trying to recall what we did at the last lesson. This is a classic example of tiger parenting undermining itself.

Regarding Sunitha's planning, the concrete advice is: avoid acceleration, *set up time for self-directed study that will not be re-assigned to other activities*, and sign-up for enrichment classes. I would like to be very specific and recommend that enrichment classes should never take up more than 30% of the time allocated for studying math outside of school. If you have signed up for a weekly 2-hour class, you should also plan for 6 hours of independent study. If you signed up for two classes in parallel, you have to make up by reserving 12 hours of independent study. Signing up for *many* classes is a proof of greed, not of commitment. The class, after all, no matter how much the instructor tries to engage all students, is mostly a passive form of learning: the student listens to the instructor, watches the instructor draw, calculate, and explain. The real commitment is represented by the independent effort and work of the student: after a class, the student researches related topics, notable problems that pertain to the lesson, a variety of problems, and, of course, spends time thinking about and struggling with a few problems of higher difficulty. During class, the instructor can only go in so much detail and depth, since the class must cover the crucial concepts in the allocated time, but after class the student should continue to research, ask questions that were not answered by the lecture, find and study more examples. Parents generally believe that the more class time they purchase, the more they help the student progress. It is not so. The class is like a guided tour through the material and represents only a starting point for independent work that enables the student to become an expert, rather than a tourist, of the area of study. Parents should avoid turning students into perpetual tourists who wander through topics and concepts!

# Independent Study

*You can know the name of a bird in all the languages of the world, but when you're finished, you'll know absolutely nothing whatever about the bird... So let's look at the bird and see what it's doing – that's what counts. I learned very early the difference between knowing the name of something and knowing something.*

– Richard Feynman

*When do we know something?* Has it ever happened to you that you purchased lots of fine ingredients and tried your hand at baking, only to be disappointed by your first few creations? If yes, then you have the experience necessary to understand that *we know something really well when we are able to create in that field.*

The real baker no longer works with a recipe book! The real baker can anticipate the result of adding a tad more milk, or of stirring in an extra spoonful of cream. He can tell if the end result will be flaky, crunchy, or moist, just by looking at the dough. He can look at two types of flour and tell you which one is better for millefeuille. He *knows* how to bake.

When students come to ask for private lessons, I see a significant discrepancy between what they believe they know and what they ac-

## INDEPENDENT STUDY

tually do know - which is why I see it is useful to explain the meaning of 'knowledge' in problem solving.

Parents often push students to take accelerated classes. At the end of the class they really believe that the student now *knows* (or should know) all the math up they have learned. If the student participates in a competition and does not come close to the parents' expectations, the conclusion is often that *more advanced* math is necessary.

For example, this is what a parent wrote to me: "I don't know which topics you are going to teach at private sessions. I listed some of them that I believe are important to increase her math ability and level: Burnside theorem, how to use Polya Index to solve symmetric counting problem, twelvefold counting, probability with infinite steps, game theory, information theory." And, while I find these topics fascinating, I still have to insist that the student be well-versed in more basic topics such as, for example, the analysis of quadratic polynomials. When I do, the immediate reply is: "But of course she knows *everything* about quadratics." Okay, I say, then can she tell me for which values of the real parameter $k$, the equation $(k-2)x^2 - 2xk + 5k =$ has two real roots both smaller than 4? Then it turns out that, knowing all the textbook information about quadratics is not sufficient to enable the student to find a reasonably comfortable strategy for solving this problem. It turns out we have to bridge the gap between having read and memorized the rules for operating the oven and the recipe and being a master baker. So, *when do we really know* quadratics? When our knowledge *enables us to be creative in answering and even posing questions* about them.

There is a big difference in ability between being able to apply some knowledge to a direct question such as 'When did the French Revolution happen?' and being sufficiently creative in this topic to be able to write a paper about 'What are the connections between the goals of the French Revolution and the Constitution of the United States' Being able to solve quadratic equations, being able to 'complete the square' are not opportunities to apply creativity to quadratic equations. When the student can respond creatively, by combining notions of polynomial analysis, to put together a strategy that can be executed elegantly, we can say that yes, she knows a whole lot about quadratics.

Therefore, the answer to the question: 'My son has an A in pre-calculus. Does he *know* pre-calculus?' is, in fact, no, if this is the only evidence we look at. He has learned some basic facts, has performed a few times repetitive patterns of application to a few common problems, and recognizes some of the vocabulary. But, in my experience, if we ask the student if he knows the binomial theorem and the answer is 'yes,' it is generally unlikely that the same student will be able to recall the theorem itself or give an example of its application. This is not a level of pre-calculus at which the student can *solve problems*, i.e. come up with original, independently derived ideas for solving. On the other hand, if I tell the parent that we have to review pre-calculus, the parent will walk away and think 'I'm not paying for *that*. Sean has passed pre-calculus already,' or 'Sean can review pre-calculus on his own.' But, you see, this is not about reviewing, it's about going up a step from one level of knowledge to another, from 'having seen Brad Pitt on the screen' to 'having had lunch with Brad Pitt,' as I sometimes tell students.

Most of the time, students are not able to take this step forward on their own. Why? Because they cannot yet make the difference between having heard of a concept and demonstrating expertise in using it in problem solving. It is more comfortable to think that, an A in algebra I means the student is an expert in algebra I. In problem solving, the difference between superficially skimming through a topic, completing basic, direct applications and *demonstrating creativity within that topic* is a difference in kind, not in degree. No amount of basic, direct applications will help ignite the creative spark. A different learning behavior and experience is needed for creative work vs. routine task completion. The less challenging the regular school homework is, the larger the rift becomes.

## INDEPENDENT STUDY

So how do we build appropriate study skills?

First of all, we must come to terms with the fact that going *back* to topics that were studied in school is an acceptable idea which will not hurt the pride of students and parents. We must realize that the difference between completing a number of direct applications (superficial study) and manipulating the concepts creatively (problem solving expertise) is yet to be bridged and, for this to happen, we may have to go back to even the simplest notions if there is valuable knowledge to be learned.

Here are a few of the most important study skills for problem solvers:

*Reading comprehension* is crucial for understanding problem statements. However, because problem statements are crafted in extremely concise and precise language, the task of comprehending all the details that play a role is often a daunting one. One must learn the mathematical vocabulary *in context*, i.e. within the theory that originated it. For example, the notion of vertex of a parabola is quite different from the notion of vertex of a triangle. Not to mention how many different meanings are given to the symbol | - it can mean: 'such that', 'divides', or be a delimiter for the absolute value notation as well as for the notation for the cardinality of a set. The student must always understand the context before attempting to figure out the meaning of the symbols or words used.

Moreover, one has to interpret not only the language but also the lack thereof. If a problem does not mention the type of triangle used, one must assume it is scalene. Or, if the statement says 'integer' one must assume that both positive and negative integer values are valid.

The language used in statements often offers clues for unraveling the problem. For instance, if we say 'Given two distinct points $A$ and $B$, what is the probability that a point that is randomly chosen is closer to $A$ than to $B$?', the words 'closer to a point than to another' hint to the fact that a perpendicular bisector of the segment $AB$ would be important in the solution. It is evident from this example that the theory must be extremely well internalized in order to allow students to make such connections, which are far from direct.

*Building up the ability to focus* and the stamina needed to maintain focus for a longer period of time has become increasingly difficult in modern times. Even so, this has to become a priority in developing good independent study skills.

*Organizing the data* is one of the most creative aspects of solving. Creating a model, diagram, table, figure, or graph can offer a hint for finding a solution. Organizing the data can help find an idea for solving or, such as in some counting problems, can actually *be* a solution. Students must be trained early to play with the data that is given.

I often observe students listing out all the possible configurations in a counting problem. Many of them list the configurations chaotically, the way they come to mind. This makes it difficult for them to ascertain that they have not omitted any case or to notice if they may have listed the same case multiple times. If the configurations are listed in some order, then it is much easier to ensure that the counting has been done correctly. Thinking up a way to organize the data can provide, however, much more than just peace of mind when counting configurations. In many counting problems, it can *suggest* an efficient counting method that had not been considered before.

I often tell young students to enact some of the word problems using stuffed animals, legos, or little cars. The word problem is a story and, by enacting it, students gain valuable insight into the requirements of the statement. For example, in the problem '*In a toy store, there are five shelves. On each shelf there are four plush animals and three cars. Someone comes and purchases two plush animals and one car. How many toys are there left in the store?*' students may take one of the following courses of action:

- Not noticing that there are five shelves, they subtract the purchases from the ware on only one shelf, obtaining the answer

## INDEPENDENT STUDY

4.

- They subtract the purchases from the ware on one shelf and multiply by five, obtaining the answer 20.

- They figure out the total number of toys: $7 \times 5 = 35$ and subtract the purchases from it, obtaining the correct answer 32.

By enacting the statement using actual toys, the student has to take the time to notice all the given data and the *process* that connects the data. The enactment *is* the solution, because *first* we have to put the toys on shelves, thereby obtaining the total, *then* we pretend someone purchases some toys, thereby executing the subtraction. Enactment is a good way to understand the order of the operations involved, as well as to observe all the details. The practice of enactment trains the skill of building a model of the sequence of operations involved and that of understanding their order.

Notice that the expression 'to figure out' actually comes from the act of creating a figure, a sketch, a visual model that helps organize the data and place the details in a context. Yet, few students have a habit of instrumenting a solution on paper.

*Building a strategy* must be the focus of any solution, regardless how simple the problem is. Simple problems should be used to develop strategies that can be scaled to handle more difficult problems. For example, the problem '*Find the smallest positive number that gives a remainder of 3 when divided by 4 and a remainder of 2 when divided by 7,*' is easily solved by students through trial and error. Almost any student will take but little time to come up with the answer 23. However, using trial and error to solve the problem '*Find the smallest positive number that gives a remainder of 18 when divided by 29 and a remainder of 7 when divided by 18,*' is not going to happen as quickly. The point of the simpler problem is to offer a simple framework for the student to build a strategy that can be scaled to the more difficult problem. The purpose of the simpler problem is not *to find the right answer*, but to find it in a specific way - by developing a strategy. When we train for problem solving, we must ask '*what was your strategy?*'

and not *'what is the right answer?'* Of course, this training process can hardly be standardized.

*Coping with difficulty* and failure is not something that can or should be eliminated from the process of solving problems. Parents often desire that students train without experiencing anything but a gentle slope of increasing difficulty that has no trace of struggle. This cannot be achieved by explaining the solution to all the problems students flag as difficult. Training should address the difficulty aspect itself: what is it that makes the problem difficult? Most likely, the inability to find a usable strategy. Therefore, students should be trained to take certain actions that push the envelope of their ability further:

- build a diagram or figure that clarifies the meaning of the statement;
- pose and attempt to solve a similar problem with smaller numbers (for example, calculate $1 + 2$ instead of $1 + 2 + \cdots + 999$ then gradually add more terms and notice what happens);
- research the words/concepts that make the statement difficult to approach;
- try to take a possible step, regardless whether it is clearly in the direction of a solution or not;

Postpone for as long as possible the following actions: looking up the official solution, google-ing the problem statement, asking someone for help.

From a training perspective, *it is more useful to spend time trying out possible approaches to a difficult problem (even without success)* than to solve several easier problems correctly.

*Making an accurate drawing* - by hand or with compass and ruler - provides insight into the construction of any geometry problem. By attempting to draw accurately, we have to figure out on the spot which configurations are possible and which are not, as well as which *order*

## INDEPENDENT STUDY

for drawing the various elements is the most appropriate. The order of drawing can help infer a lot of information useful for solving - more specifically, useful for generating an idea. It should be noted that accurate drawings with compass and ruler are required by the USA(J)MO exams.

Unfortunately, due to the need for standardization in the grading process, multiple choice problem statements include an example drawing. Moreover, also because of the difficulty of grading and providing feedback, skills for technical drawing are no longer taught. Students who rely on drawings already provided by the examiner are like people who eat pastry and are asked to provide a recipe: just by eating it one cannot figure out which ingredients to use and, more importantly, *which order they must be used in*. Of course, the beginner's picture is obscured by the availability of the drawing provided by the statement: we think such a drawing is *supposed* to be provided, we think it is a waste of time to re-do it. Again, it is not the final aspect of the drawing that helps, but the process of *assembling it from its components in the order of drawing that we are constrained to by the statement* that provides insight into the problem.

*Aim for an elegant, efficient strategy* for each and every problem, even the easy ones. Even if you can rapidly find a correct answer to the easy problem, remember it is not the correct answer, but the scalable strategy you are seeking: a strategy that will adapt to larger numbers, larger number of operands, to problems which are similar but much more difficult to solve.

# A Training Methodology

*If we are facing in the right direction, all we have to do is keep on walking.*

– Buddhist saying

Which training method works best? For those who expect me to reveal the best extracurricular program or the best instructors, I must disappoint: the best method is to study independently. However, even independent study is subject to simplistic approaches which may render it very inefficient. It is common practice for parents or instructors to obtain a past test paper, ask students to solve it in the allocated time, and grade it. While this in itself can be a valuable teaching tool more often than not the answers are simply compared to the answer key and a brief, often cursory look at the official solutions follows. The student declares 'I got it now,' and starts working on the next practice set. This training method is only marginally useful and I do not recommend it as it can be a waste of good training material.

Solving many problems is necessary but not sufficient. Students must also commit time towards learning theory and connecting problems to the theoretical facts they are built on. Instead, students read existing solutions in an attempt to figure out 'how it was done.' Often, in doing so, they focus on the actual execution steps and not on the reasoning that led to those steps. In this manner, students learn

*TRAINING*

theory in a stunted, incomplete way - as a result, their capacity to innovate remains limited to imitation. When studying past test papers, students have to combine quantity (i.e. number of problems solved) with quality (i.e. the level of detail of their study.)

**How should we study problems?**

Training is different from competing! When we train, we should not focus from the start on the speed at which we solve problems. The speed will increase naturally as our experience increases. More specifically, increasing the speed of solving is one of the reasons I tell students to try to refine their solutions even after having obtained the answer. Now that they have a solution, they have to think if there is anything they can do to *optimize* it either by simplifying parts of it or by adopting an altogether different strategy. By comparing different strategies for the same problem, either obtained alone or by studying other people's work, we become more and more efficient as we hone the ability to select a shorter, simpler strategy sooner in the process of solving. *Improving the speed at which we solve is a long term goal, not a short term goal!*

Do not solve past problem sets with a timer, unless you are really giving yourself a test. This should occur at most every two months. Otherwise, give yourself plenty of time to study each problem in detail. Each problem, even ones that were answered correctly, should be analyzed with the following questions in mind:

- What is the underlying theory?
- Are there several ways of solving it?
- Which one of these is the most efficient?
- Which is the safest method of solving?
- Are there special execution details?
- Can the execution of the solution be simplified?

*TRAINING*

- Can we imagine other problems based on a similar statement?

These questions are crucial for training, and are useful whether *or not* the student has obtained the correct answer to a problem. This is because a student can produce a correct answer in a manner that does not help develop creative solving skills. These include:

- lucky guesses,
- plugging in random numbers (yes, students do this!)
- plugging answer choices back into the statement,
- using logic to eliminate as many answer choices as possible,
- using logic to eliminate some choices and then guessing,
- using brute force (such as, for example, making a long list of all possible numeric scenarios.)

Although some of these ways of obtaining a correct answer may be useful during a competition, they do not, for the most part, help with training. As we will explain later on, most of these ways of selecting an answer - we cannot call them solving - are actually harmful.

Training should focus on improving the solving methods students use - a task that is related, but is not identical, to that of maximizing the number of correct answers on practice tests.

A useful solution must be *based on proof*. A problem usually describes a system in which some information is available (the *known data*). The student must start from known data and progress step by step towards a better knowledge of the system by using logical connections among mathematical facts that have been proven beforehand (the *theory*). Thus, the student generates *derived data* in a small scale process of data mining, thereby enhancing their knowledge of the system. This additional knowledge is either sufficient to answer

79

the question asked, or must be used as a starting point for a yet more advanced stage of solving.

The proof that forms the solution must be written out informally, yet clearly, in a language that enables the reader to follow the logical inferences used by the solver.

**Here are the steps of *proof-based* solving:**

1. Read the statement, carefully observing all the details.

2. Sort the data into two categories: *given* and *needed*.

3. Mine for the needed data using logical reasoning that is based on proven facts.

4. Use the data collected in step 3 to take a *creative step* that generates an idea for a solution.

5. Execute the operations that embody the idea.

6. Verify that the answer obtained matches the requirements of the statement.

Since an official solution presents only step 5, students obtain little of no benefit from 'looking up the solution' for problems they are not able to solve independently. When they say "I understand how it's done," they mean "I understand all the steps described in the solution." That is, they understand what formulas have been applied and how the operations (multiplication, addition, powers, etc.) have been performed. But the crucial step is step 4! Obviously, no written solution explains how the solver is to get *the idea*. To make proper use of an official solution the student must use the description in order to *figure out steps 1-4 and how they lead to the operations listed in step 5*. The student must work with the official solution much more than by just looking it up and deciding that the operations 'make sense.'

Watching 'how it is done' (say, from an instructional video) is only a little better than skimming through the official solution. Problem solving is cannot be learned it by watching someone else solve problems.

**How should we read the statement and prepare for solving?**

- Read the statement of the problem until you start to understand it. This may succeed only partially. You may revisit this step later.

- Try the exercise described in the statement on a smaller scale. For example, if they say add 130 numbers of some type, add just 2 numbers.) Try out the process described in the statement on smaller scale examples, until you feel comfortable saying that you really understand the statement.

- For geometry, *make your own drawing*, by hand, on paper. Do not incorporate any symmetry that is not mentioned in the problem. For example, if the statement does not say the triangle is isosceles, do not draw it so. Make a special effort to keep unwanted symmetries out of your drawing. Read the problem again and *redo your drawing*, if needed. Personally, I do not find that accurate drawing with ruler and compass is more helpful than a reasonably good hand drawing - but, of course, drawing skills must be part of training.

- For word problems, make a simple diagram: a timeline, a number line, a clockface, a table, etc.

- For algebra, examine the expressions given and try to figure out first if the numbers you work with (unknowns, as well as coefficients) are integer, rational, real, or may have imaginary parts. If the problem is dealing with integers only, you may have to solve a number theoretical problem and not a purely algebraic one. Make a note to yourself that you may have to solve *Diophantine*, not algebraic equations. The nature of the numbers involved is already a clue for finding an idea.

*TRAINING*

- Sort the data into *given* and *needed*.

- Write the data out separately on your paper, beside a drawing, if you have one. The data should be listed clearly and unencumbered by text. If needed, make up variable names for some of the data: for instance, denote the number of miles with $D$ and the time with $T$.

**How not to read the statement**

- Do not read the statement in haste. No detail should be overlooked. Make sure you understand the grammatical structure of the sentences: which numbers are you asked to calculate the difference of? which of the numbers is supposed to have three digits? etc. Take the following problem, for example:

    Anna subtracted from the largest whole number with all different digits the third of the smallest whole number with all different digits . Which number did she obtain?

    Think: Anna subtracts this from that. Now match 'this' and 'that' to the numbers described by the statement. 'This' is 'the third of the smallest whole number with all different digits,' while 'that' is the largest whole number with all different digits. Compute each of these numbers and perform the subtraction.

- Do not skip a problem with a long statement just because it seems like a lot of work. Comprehending the statement is in some cases the most work a long statement will require.

- Do not try to 'find the answer' as the next step in the process. Between reading the statement and finding the answer there is often a lot of work that must be done.

- Do not use the drawing provided in the exam paper. You need a drawing that you complete yourself by following the instructions in the statement. As you draw in more elements, you will start noticing which parts of the drawing depend on others. Except for the simplest drawings, there is often a specific order in which it is better to draw the details. By attempting to make as accurate a drawing as possible, this order will become apparent - it often provides a key to the solution. Making one drawing is often not sufficient. Do not hesitate to make more drawings - the effort is well worth the benefit! By improving the drawing you also improve your knowledge of its geometry.

**How can we find an idea?**

Ideas can only be generated when we already know a lot about the system. Theory is a pre-requisite, not an option, and it should be treated as a set of problems to solve. For each theory topic the student must understand:

- what problem it is trying to solve;

  For example, when students learn the remainder theorem they are told that when dividing a polynomial $P(x)$ by $(x - a)$, the remainder is equal to $P(a)$. Subsequently, many students incorrectly assume that, if the remainder is $P(a)$ when dividing by $(x - a)$ and $P(b)$ when dividing by $(x - b)$ then the remainder is $P(a) \times P(b)$ when dividing by $(x - a) \times (x - b)$. They remember from number theory that, when the remainder of a product of numbers is equal to the product of the remainders. However, the division problem for polynomials is to divide the same polynomial by different divisors, while the number theoretical fact refers to dividing different numbers by the same divisor.

- how this theoretical fact is connected to other facts;

  For example, the angle bisector theorem, the power of a point theorem, Ptolemy's theorem can all be proven using the similarity of triangles. Since all these theorems establish connections among ratios of segment lengths, we conclude that, in general similarity of figures is a useful tool in problems that ask for ratios of lengths.

- if this fact can be generalized;

For example, if we look at combinations as binomial coefficients, can one think of a similar concept for trinomial coefficients? If so, is the interpretation of a combination as the number of ways to select objects from a set going to be transferred to the new concept? If not, then why not?

- if the fact can be re-purposed;

Many students think that if the rule for divisibility by 3 says "For a number to be divisible by 3, the sum of its digits must be divisible by 3," then the rule for divisibility by 7 should say "For a number to be divisible by 7, the sum of its digits must be divisible by 7." This is because the student knows the rule but does not know *why* it works. At a loss when facing a problem about divisibility by 7, the student erroneously tries to simply re-purpose the rule for divisibility by 3.

- what range of applicability the fact has;

Students memorize the fact that irrational roots of polynomials occur in 'conjugate' pairs (if $2+\sqrt{3}$ is a root, so is $2-\sqrt{3}$) but only rarely know that this happens if and only if the coefficients of the polynomial are rational numbers. This happens, for the most part, because students learn this fact *by reading solutions or by watching instructors solve problems*. The written or presented solution simply applies the theorem without explaining *which pre-requisites must be fulfilled for its application*.

### What are the solving behaviors we should avoid?

Do not solve a problem *by analogy with another problem* similar to it. Subtle differences in the wording may result in very different solving strategies. Always solve a problem from scratch.

Information that is not mentioned in the statement cannot be assumed true without a proof. A simple example is the following traditional problem:

Two men approach a river. There is a boat tied to one of the banks. The boat is small and can only carry a single person. Both men crossed the river without getting wet. How could they have done this?

As they read the problem, most people will automatically assume the men are walking *together*. However, this fact was never mentioned in the statement. The answer is that the men approached the river from opposite directions. It is very common for our brains to embellish the information provided in a story. The brain feels the need to complete missing details and does so based on our previous experiences.

Defective solutions for more advanced problems can often be traced back to similar lack of discipline in managing information. Students assume some angle is a right angle because 'it looks like that,' some meeting takes place 'in the middle,' some numbers are whole, etc.

Finding an idea is often related directly to practicing discipline in managing the given data. For example, just because 4 points *can* form the vertices of a square, doesn't mean that the figure they belong to is a square. It could be a circle or a set of circular arcs, to name a few possibilities:

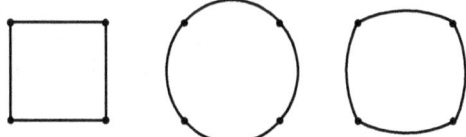

Yet I have seen students often assume without proof that the points are on a square or a circle.

Do not re-purpose rules! Do not use the Pythagorean Theorem if you do not know for sure that the triangle has a right angle. Do not assume that if 'even plus even equals even' it is also true that 'odd plus odd equals odd.'

TRAINING

**How do we execute the computational steps required in order to finalize the solution?**

Often times, we believe it is obvious that, once we have the idea for a solution, the rest of the process should be child's play. In reality, execution typically requires employing shortcuts that belong to the toolbox of an experienced problem solver. When students watch an instructor solve, it is not sufficient to say 'I understand what they did.' They must also ask 'What would I have done?'

As a common example, we encounter problems that probe if the knowledge of the quadratic equation extends past the quadratic formula. The simplest of these asks for the sum of the roots of an equation such as:

$$4x^2 - 13x + 17$$

The student who knows Viète's relations will immediately answer 13/4. The student who does not will attempt to find the roots and add them. Notice that the equation has negative discriminant and the roots are complex numbers.

Executing computational steps often requires taking imaginative shortcuts or using the order of operations in a creative manner. Simplifying expressions should not be seen as a repetitive process of applying the order of operations. Instead, students should practice processing expressions without a calculator in order to strengthen their capacity to put together a plan for executing operations in the most favorable order. To achieve this, they must practice their observation skills (observe patterns that may lead to the application of algebraic identities), their ability to predict in advance the effect of a specific order of operations, and their speed in making decisions that narrow down the various options for computation.

**How do we structure the math content itself?**

Students ask me often: 'How should I divide my time between geometry, algebra, counting, and number theory?' As well, parents prefer to see the content taught topic by topic. Even our own training workbooks are organized by topic, albeit not very strictly. However, I must be the contrarian again and say that, for the most part, it is more efficient to work on mixed review. I divide the work in 40% topic-based study and 60% mixed review problem solving. The mixed review allows students to keep all the concepts they learned constantly active and to improve their skills across the entire breadth of the syllabus. Moreover, a lot of the more difficult problems require connections among different topics (geometry and counting, algebra and geometry, number theory and algebra, etc.) and it is not clear how to categorize them.

Lectures should be turned into *mini-lectures* that briefly explain the central concept. The mini-lecture should be practical, not expository: the lecturer should not build a theoretical framework expecting that the students will investigate its applicability and connections to other concepts on their own. The lecturer must cater to a short attention span and start by advertising the framework: what is it good for and how does it make certain solutions simpler? Today's student is pragmatic and must understand the benefit of a method before internalizing it.

Mini-lectures should be followed by examples as well as counter-examples of applicability, if possible in the form of similar statements that are within vs. outside the range of applicability of the concept. This will keep students alert to the danger of imitating solutions. This approach uses a 'good guy vs bad guy' stereotype will make the topic look less static and will keep the students focused.

The mini-lecture should be proof-based - i.e. they must rely on an informal proof. *Proof-based* mini-lectures are not isolated chunks of knowledge but have (or should have) a number of links to previously studied material as well as to future topics.

In designing a sequence of lecture, one should use *mindful repetition*. Students cannot retain 100% of the material presented to them but through continued repetition the retention will increase. Repetition is boring. Moreover, students resist repetition by forming mechanical patterns of execution that they injudiciously apply without further analysis. Mindful repetition is a technique for learning in context:

1. by applying the concept to contexts that differ substantially,
2. by applying the concept to contexts that test the limits of its range of applicability,
3. by applying the concept to contexts that are outside the range of applicability,
4. by applying the concept to contexts that 'mimic' the typical context,
5. or by applying the concept 'backwards', i.e. from final value to values of the parameters.

These applications have to be:

1. explained by the instructor,
2. explained by the students in their own words,
3. debated by the group,
4. and re-explained by the students in their own words.

The student must not repeat what has been explained, but produce other, new examples, as a first step in creating solutions.

**Counting example 1:** In how many ways can Amira place a red ball and a blue ball on Lila and Dina's desks?

In two ways: the red ball on Lila's desk or the red ball on Dina's desk.

**Counting example 2:** In how many ways can Amira place a red envelope and a blue envelope on Lila's and Dina's desks?

# TRAINING

There are four ways of placing an envelope on Dina's desk: the red envelope facing up on Dina's desk, the red envelope facing down on Dina's desk, the blue envelope facing up on Dina's desk, and the blue envelope facing down on Dina's desk. For each of these, there are two ways for placing the other envelope on Lila's desk. The total number of ways is $4 \times 2 = 8$.

Worksheets should be in *random difficulty order*. In this way, the student will not have the opportunity to select which problems to attempt and which to ignore without at least giving them a try. This approach will also train the student to estimate more accurately the time it may take to complete a solution, a very useful skill in high level competitions such as the AIME. On competitions such as the AMC it is a useful skill to be able to choose the problems one intends to solve and leave a few, perhaps more difficult, problems unsolved for later. But how does the student know which problems to choose, especially under time pressure? Training this ability is another reason for working on problems in random difficulty order - the student is forced to comprehend all the statements and assess the difficulty of the problem. Often, a little bit of work must be spent on trying out some ideas that will not succeed.

*Mathematical vocabulary and notation* should be used at all times and introduced as early as possible. Problem statements are very concise and crafted to convey a lot of information in the most precise manner possible. There are specific ways in which statements are written and they all use standard notations and vocabulary. Expectations for the problem to be stated using simple vocabulary and without notation can only be met if the entire mathematical language is degraded, resulting in longer, more inefficient, and *difficult to remember* statements. Therefore, students should be trained to pay attention to the smallest details in the language of the statement. As an example, compare the following counting problems:

**Example 1:**

How many positive factors does 952 have? Answer 16.

TRAINING

**Example 2:**

How many integer factors does 952 have? Answer 32.

As the number of factors can only be a whole number, both problems are discrete counting problems. *Positive factors* refers to the number of distinct positive integers that divide the number exactly: $1, 2, 4, 7, 8$, etc., while *integer factors* refers to the number of factors both positive and negative: $\pm 1, \pm 2$, etc. Notice how a subtle difference in describing the objects to be counted makes a considerable difference in the answer.

Problem statements sometimes omit information that they deem implicit. For example, the problem statement:

"What is the next term in the sequence: $2, 4, 8, \ldots$?"

actually means:

"Assuming the observable pattern continues, what is the next term in the sequence: $2, 4, 8, \ldots$?"

If the assumption that the pattern continues is removed, then it is impossible to tell with certainty what the next term will be. Sometimes, however, problem statements omit this crucial information. This leads students to believe, erroneously, that the observable pattern *always* continues. Notice how, for example, the sequence defined recursively as $a_{k+1} = a_k + 2k$ and $a_1 = 2$ continues with the terms 4 and 8 but the next term is 14, not 16. Students cannot rely on computing the first three terms and relying on the observable pattern (powers of 2). They should, instead, analyze the recurrence relation and conclude that it is neither arithmetic, nor geometric. After this conclusion, it is obvious that the sequence does not consist of powers of 2 and, therefore, cannot be solved by assuming the observable pattern continues. In the case of this last problem, the recurrence relation is a pattern in itself, but it is not the same pattern as the powers of 2 that can be noticed at first glance.

Using standard mathematical vocabulary from the earliest years of education will ensure that students learn this language gradually and in connection to its context. Also, it ensures that the students are exposed to sufficient repetition and to variations of the context.

*TRAINING*

# Mathematical Competitions in the US

*Nothing any good isn't hard.*

– F. Scott Fitzgerald, A Life in Letters

In 2016, more than one hundred thousand students participated in the American Mathematics Competition in the USA. What drove them to it and what did they get out of this experience?

The vast majority of students who participate in mathematics competitions do so because they are advised to do so by their parents. More and more schools, both public and private, offer the opportunity for participation under the constant, strong pressure applied by parents. Parents engage students in mathematical competitions for a variety of practical reasons. Perhaps the most important of these is the ever increasing advantage that successful participation in math competitions confers in the process of admission to college - parents definitely see math competitions as the gateway to the Ivy League schools and other top tier universities. Lately however, parents have become even more concerned about the job market and their offspring's fitness as candidates for well paid positions - parents are no strangers to the fact that mathematically strong candidates are in high demand in fields such as finance, risk analysis, and computer engineering. Success in math competitions reaches beyond college, as far as job applications and aspirations for a high personal income are concerned.

One time, on a last day of class, I asked Liam, a student of mine who had participated in Math Leagues at the State level and had qualified to participate in the National finals: 'So, when are you going to the Nationals?' And Liam replied: 'I'm not going. My parents don't want to send me. I begged them to, but they said they're not sending me because I don't have any chance of winning.' His parents did not understand that *a math competition is not a final product of learning* but a learning opportunity. By participating in competitions students train and exercise a lot of important skills: focusing over an extended period of time, battling difficulty, generating creative responses in a limited time, while also getting unique feedback about how they perform under stress and how their performance could be improved. The goal of a competition is to engage students in participating, not in winning. Winning a competition is an extremely difficult task that requires not only exceptional skills but also a bit of luck on the day of the exam. Students who participate regularly in math competitions know that it is possible to do very well on one of them and then less well on another one, even soon after. I had students who performed extremely well on AIME and were invited to the USAJMO, but did not manage to obtain a prize on Math Kangaroo. Liam's classmate Sophia won first prize nationally on Math Kangaroo and, in her own words, 'did very, very bad on Math Leagues', - such stories are not uncommon.

Unlike athletic competitions, math competitions require 'getting an idea' - or, in fact, several good ideas - an event that has much more variable an outcome than, say, ice skating. An Olympic ice skater who has won the gold medal is holding the final product of her training. She needs a little bit of luck, but she does not need to come up with new routines in the middle of the competition. A math olympian's medal is only a stage in her training and is only partially relevant to her future productivity as a mathematician. Participation in a math competition is part of a process of learning. All students should take advantage of every opportunity to compete because it helps them sharpen their

## USA MATH COMPETITIONS

skills, not because they have a chance to win!

In the USA, the most prestigious contest is the American Mathematics Competition (AMC), run by the Mathematical Association of America (MAA). The AMC has a middle school competition - AMC 8 - which takes place in the fall. The AMC 8 is a single stage competition. Results on AMC 8 are not useful for college admission - it is an event that allows students to get their feet wet and to gather feedback about their own performance in a timed math competition. Students may begin to take AMC 8 in grade 6 and they can take it each year until grade 8. The high school level competitions are AMC 10 and AMC 12. Students may take any of these competitions ealier than recommended but the last time they can take AMC 10 is in grade 10. Similarly, the last time they can take AMC 12 is in grade 12.

There are two dates for each of these exams, usually both in February of each year. A student may take two of these exams each year in any combination (both AMC 10 A and B, or AMC 10 A and AMC 12 B, etc.). The highest score obtained in any one year may qualify a student for the American Invitational Mathematics Examination (AIME). There are two dates for AIME as well, AIME I and AIME II, but a student is not allowed to take both. AIME II is generally provided only if, due to some accident, a school cannot provide AIME I, or a student has a good reason to be absent that day.

The scores on AIME and AMC 10/12 are combined using a formula to produce the USAMO index, a score that may qualify a student to participate in the United States of America Mathematical Olympiad, of which there are two flavors: the junior competition (USAJMO) for students who qualify based on their AMC 10 score and the senior competition (USAMO) for students who qualify based on their AMC 12 score. At the USAMO, a smaller group of students in selected to candidate for membership in the USA team that will represent the country at the International Mathematics Olympiad (IMO), a very prestigious competition.

### Everything You Need to Know About the AMC Exams

The AMC exams are based on mathematical knowledge gained in school as well as outside of school, since it is a test for students who have a special interest in mathematics and who love to spend time learning more math and discovering new ways to solve. These tests reward those students who:

- are sufficiently interested in math to learn on their own, outside of basic school requirements;

- are inquisitive and readily spend time figuring out improvements to proofs and solutions, or researching in more depth the facts that they have learned in school;

- solve problems using a proof-based process (not by 'trial and error', 'plugging the answers back in', 'elimination of some answers', or 'guessing');

- are capable of mathematical creativity and can rapidly modify a proof they know, or devise on the spot an efficient method of counting, or other technique;

- can focus and work extremely well under time pressure.

A knowledge of calculus is not required for any of the AMC exams, nor for any high school level mathematical competition leading to the IMO. The main areas of knowledge needed are arithmetic, number theory, algebra, geometry, counting and probability. You can find a detailed syllabus in the Appendix.

AMC exams are in multiple choice format, while AIME is in grid-in response format. Written informal proofs are required on USA(J)MO and higher. However, even for the AMC and AIME levels, problems are devised so as to require a proof-based approach. On many occasions the examiners go to some lengths to make sure no hint is given by the answer choices.

AMC problems are very professionally crafted and stated. The student must be familiar with high level mathematical language and terminology that may or may not have been used in the school textbook. The student who participates in AMC has to prove readiness for a scientific profession in which writing and understanding specifications are basic skills.

AMC problems are not direct applications of the material studied in school or in extracurricular classes. Exceptions are the few very simple exercises at the start, which are essentially warm-ups, designed to help the student focus. Otherwise, the problems range from challenging applications, to problems that require the ad-hoc extension of a mathematical fact to a situation it does not cover, i.e. the spontaneous creation of a new mathematical concept or procedure.

On all these exams, the problems are listed in somewhat increasing difficulty order: problem # 1 is definitely much easier than problem # 25, but along the way, it is often possible for a problem to be easier than the preceding one.

# All About AMC 8

AMC 8 is a 25-question exam that must be completed in 45 minutes. The student is required to find clever strategies to solve the problems and must solve at quite a brisk pace. Generally, problems 18 through 25 require creative response.

As many as one third of the problems are based on Euclidean geometry in two and three dimensions of medium complexity: parallel lines, rectangular shapes and solids, dissections of plane figures, area and circumference of circles, area and perimeter of triangles and quadrilaterals, angle chasing in plane figures, simple properties of angle bisectors and perpendicular bisectors, simple cases of similarity, special triangles. Geometry problems may also occur in combination with counting (sum of interior angles of a convex polygon, number of diagonals of a convex polygon), in combination with number theory (figures with integer sides, triangle inequalities, connection to Fibonacci sequences, etc.), or in combination with algebra (parametrizing elements of figures, setting and solving equations.)

Another third of the problems may be number theoretical. At the time of writing, we have seen an increased occurrence of problems based on linear Diophantine equations and factorable Diophantine equations. However, in general, we see problems based on: Euclidean division, divisibility, bases of numeration, modular arithmetic, sequences, perfect squares and cubes.

Counting and probability is another major topic on AMC 8 and math contests in general. Problems posed on this exam are based on elementary counting techniques, yet, many of them cannot be solved by direct application of a standard technique, and require some ad-hoc counting method.

Problems based on pre-algebra and algebra are, for the most part, among the easier problems on the exam. The expectation is that students will have an excellent number sense that enables them to use algebraic identities and the order of operations in creative ways.

There is no penalty for guessing on AMC 8. Each correct answer is rewarded with a point, and incorrect or missing answers do not contribute to the score.

What does a score on AMC 8 mean? After receiving the scores, parents may want to look up the ACM statistics page of the MAA, where a statistical breakdown of the scores obtained nationally becomes available once the scores are sent to schools. Using this breakdown, students and parents can find out how their performance on the exam compares to that of other students, on a scale that far exceeds the micro cosmos of their own school.

Scores below 10 points indicate that the student does not master the regular school math and will not perform very well on standardized tests (SAT, ACT, PSAT), without improving their math study habits.

It is unlikely that the student will achieve a consistent A in math through high school, more specifically in calculus class (which depends heavily on pre-calculus and geometry,) and is generally not positioned for success in an engineering profession. This is a statistical view, not a personalized analysis.

In my experience, most students do know how to execute the steps for problems such as this one:

> Find $x$:
> $$\frac{4}{x} = \frac{16}{128}$$

but may not be able to solve a similar problem that requires the same math concepts such as:

> In the set of proportional numbers $4, 16, 128$, and $x$, which of the following is a possible value for $x$?

This points to the regrettable fact that the student does not understand the the meaning of the computation involved, but just learned the mechanical execution of steps necessary to pass school quizzes, tests, and benchmarks.

Students who are otherwise 'good at math' in school may score low on AMC 8 because they lack problem solving skills. If the exercise is not direct and simple enough to allow them to form a solution procedure instantly, they are at a loss as to what steps to take and they resort to fallacious processes such as: wishful thinking ('it looks larger than the other, so it should be 6'), guessing (which the student often calls 'guess-and-check' although the checking part is missing), elimination (although this method may lead to a correct answer, it is not a problem *solving* process because in the absence of answer choices the student would have no idea what the answer might be), or plugging answer choices back into the problem (same as elimination, this is not an *enabling* strategy).

Scores between 15 and 20 generally show that the student is somewhat inquisitive and able to apply some of the math learned in school to scenarios that require some imagination or experimentation. It is more difficult to model the mathematical ability of this student. In the best case, the student is actually interested in math, is participating in a contest for the first time and, though able and creative, takes more time to figure out each problem than is allocated by the rules of the competition. In other cases, the student has prepared by studying past exam papers or by taking extracurricular classes - however, neither the efficiency nor the creativity required have developed to a sufficient degree. Other times, this score shows that the student employs solving techniques that have 'maxed out', such as listing all possibilities. Often, scores in this range motivate students to engage more fully in the study of math, especially after they realize what a powerful combination of skills they need to develop. Parents and students realize now that the path from an A in school math to an AMC score that ranks in the top 25% nationally involves more than spending an extra hour a week on problem solving.

Students who obtain scores above 20 have, for the most part, a proof-based approach to problem solving and sufficiently well trained skills for solving under time pressure. In my experience, students who have scored 24 or 25 on AMC 8 have also qualified for the AIME in the same academic year, as well as throughout high school. Most of them have later been admitted to prestigious universities such as MIT, Caltech, Columbia, and similar.

# All About AMC 10

AMC 10 is a 25-question exam that must be completed in 75 minutes. The first ten problems are relatively easy applications of school math involving pre-algebra (ratios, proportions, percentages, rates), algebra I, middle school level geometry, elementary counting, and some astuteness problems. Middle school math is generally sufficient for solving the first ten problems!

Problems between #11 and #20 are more difficult and require familiarity with a number of extracurricular concepts, albeit at a level compatible with middle school (bases of numeration, modular arithmetic, Diophantine equations), an advanced mathematical vocabulary and good comprehension of mathematical statements, good computational skills, and some amount of creativity.

The final 4-5 problems consist of rather difficult solving situations that put the student's mastery and creativity to the test. A proof-based approach is essential. Lately, the last 6-8 problems have not been listed in increasing difficulty order - this is a new trend in the format of the exam.

Each correct answer receives 6 points, while each skipped answer receives 1.5 points. Incorrect answers receive 0 points. In this respect, there is a penalty for guessing, as the AMC has the goal of promoting, to the extent possible, those students who reason out the solution in a *proof-based* manner.

As far as the syllabus goes, the AMC 10 exam may include problems based on algebra II and trigonometry such as functions, logarithms, trigonometric equations, complex numbers, polynomials, as well as on extracurricular topics such as number theory, special topics of geometry, or advanced counting methods. All AMC 10 problems have solutions that are not based on calculus!

Recently, some 1,100 students received invitations to participate in the AIME as their score on AMC 10 ranked in the top 2.5% of a total of around 70,000 scores. This invitation is highly regarded by most colleges and universities during the application process, especially if it happens consistently over time. Qualifying once and then never again is not as good: the student may have been lucky one time, or may have lost interest in math later on.

I often receive questions from parents and students who want to maximize the chances of receiving the AIME invitation. One day, a mother sent me the following very typical question: 'The AIME qualifying score is not out yet, however she is kind of scared that she might not qualify with a score of 113. At this point she really wants to qualify for AIME and so for this is it better to take AMC 10 B or AMC 12 B? There are lot of opinions out there on the Web, we would like to seek your advice.' Here is what I answered:

*The more desperate we are to get some result, the more elusive it tends to become. Put all your capacity into solving the problems, not into thinking about the score or the time. I think that, if she focuses just on the math, on getting as many of the problems done, on the thrill of having a good idea occur magically in the nick of time, that would be the best attitude and the one that will bring her the best score. Being desperate about the score itself will not help her make good decisions. If she goes there for the love of math, she will get the best possible score she can - guaranteed.*

# All About AMC 12

AMC 12 is a 25-question exam that must be completed in 75 minutes. AMC 10 and AMC 12 exam papers share a subset of problems. Some of the medium difficulty AMC 10 problems appear on AMC 12 among the problems in the 'easy' section, while some of the more difficult AMC 10 problems also appear on AMC 12 as medium difficulty problems. The last few problems on AMC 12 are typically very difficult. Nevertheless, each year, there are around five to ten students in the USA who manage to achieve a perfect score on this exam.

Each correct answer receives 6 points, while each skipped answer receives 1.5 points. Incorrect answers receive 0 points. In this respect, there is a penalty for guessing, as the AMC has the goal of promoting, to the extent possible, only the students who reason out the solution in a *proof-based* manner.

For a recent AMC 12, from a total of approximately 73,000 students nationally, about 430 earned a place on the distinguished honor roll - a benchmark recognizing the top 1% of the scores. The number of students invited to the AIME was around 1,100, recognizing the top 2.5% of all scores.

AMC 12 is more difficult than AMC 10. The vocabulary used in AMC 12 problem statements is advanced mathematical vocabulary that may or may not be used in textbooks, depending on where the test is taken. Some states, such as California, prefer to explain a concept in many plain words rather than introduce a new word that would express the meaning in a concise manner. For example, instead of *irreducible*, a California textbook would say *in its most simplified form*.

To be mentioned on the Distinguished Honor Roll is an exceptional achievement that will bring a great additional weight to a college application. Looking at the numbers, it is clear that a college should aggressively go for anyone who has received an invitation for the AIME - other are one of a group of 1,000 top math students in the nation. As I mentioned before, several consecutive invitations to the AIME constitute a particularly strong record of achievement.

# All About AIME

AIME consists of 15 questions to be solved in 3 hours. The questions are constructed so as to admit whole number answers ranging from 000 to 999 - and, yes, there have been answers of 000 on some of the problems. The three digit answers must be entered in grid-in form.

Each correct answer receives 10 points, while skipped or incorrect answers do not receive any points. On the AIME, there is no penalty for guessing - however, it is not a multiple choice exam, which makes guessing an extremely poor strategy.

The AIME score and the AMC 10 or AMC 12 qualifying score are added together to form the USA(J)MO index, a combined score that opens the gate to the next level of the competition: the USA

Mathematical Olympiad. If a student qualifies from two different exams, the higher score is used to calculate the index. Students whose index is computed based on an AMC 10 score might be invited to participate in USAJMO, while students who qualified via AMC 12 might be invited to write the USAMO exam.

Yearly, about 500 students participate in the USA(J)MO exams, roughly half of them in the junior group and the other half in the senior group.

Questions on the AIME reflect the fact that participants may come from age groups ranging from middle school to junior year of high school. There is a sufficient number of problems based on elementary facts to give younger students the chance of obtaining a qualifying index. There is also a number of problems based on advanced concepts such as: trigonometry, complex numbers, polynomials, conic sections, etc., without, however, any use of calculus.

Contrary to what most students and parents tend to believe, most of the problems on the AIME do not require very advanced extracurricular topics, but do require a very intimate knowledge of all the elementary facts. The principal topics on the AIME are: Euclidean geometry in 2-D and 3-D, number theory, counting and probability, and algebra.

# Participating in AIME via USAMTS

For the last 27 years, the National Security Agency (NSA) has sponsored the USA Mathematical Talent Search competition. The USAMTS offers three rounds of papers in October, November, and January of each academic year. Registration is free and students may download each exam paper as soon as it becomes available. For each round of problems, students have approximately one month to send in fully written out solutions based on informal proof. Each student must work individually and independently on the problem set. The problems posed on the USAMTS are generally quite difficult. Each student submission is graded by hand for correctness and writing style using a score from 0 to 5. Each year, a number of students who accumulate good scores through all three rounds receive an invitation to the AIME. Therefore, the USAMTS is an alternate way to be considered for participation in the USA Mathematical Olympiad. In my experience, students who have qualified through the USAMTS have also qualified through the AMC exams. Awards and good ranking on the USAMTS are very valuable assets at college admission time!

Both the AMC and the USAMTS share students' scores with colleges and employment agencies only with permission. Be sure to check the permission box if you would like your score shared!

# Appendix - Syllabus for Competitive Mathematics

This list reflects my experience of teaching competitive mathematics. It is not a list of *all* the topics that should be studied but a list of the special skills and concepts that problem solvers need to practice outside of, and in addition to regular school.

## Grades 1-2

Pre-requisites: addition and subtraction

- reading comprehension skills and familiarity with the language of problem statements;
- ability to infer the purpose of a variety of puzzles: visual, rule-based operations, sequences with a variety of patterns, vocabulary-based, etc.
- how we write numbers (*decimal system*) vs. how others write numbers (*e.g. Roman numerals*);
- digits vs. numbers;
- even and odd numbers;
- palindromes;
- sum, difference;
- cryptarithms (arithmetic puzzles) with addition/subtraction operations;
- how carry-over and borrowing generate clues for solving cryptarithms (arithmetic puzzles);
- pairs, double, triple;
- ability to draw a clockface;
- ability to draw regular polygons based on the clockface;
- representing time using numbers;
- converting from 12-hour time to 24-hour (military) time and conversely;
- calculating sums and differences of times (both 12-hour clock and military time);

- drawing a timeline and representing time instances on it (**now**, past, and future);
- making a weekly and monthly calendar;
- using shortcuts in the calendar;
- drawing maps and itineraries from a description (specification);
- building concrete models for unknown quantities using a box model;
- optimizing simple operations by keeping the numbers small;
- solving simple equations using a box model;
- solving puzzles with operations;
- counting the number of objects in an enumeration, both, one, or no ends included;
- counting the number of objects in a circular arrangement;
- observing and using symmetry in counting;
- consecutive numbers, consecutive even/odd numbers;
- solving visual puzzles based on: reflections, rotations, or rule-based patterning;
- practice drawing segments, circles and simple figures;

If using our series of workbooks "Competitive Mathematics for Gifted Students", the recommended order of study is:

1. Practice Observation and Logic - level 1
2. Practice Counting - level 1
3. Practice Arithmetic - level 1
4. Practice Operations - level 1

# Grades 3-4

Pre-requisites: multiplication of multi-digit numbers, division of multidigit numbers with remainder

- definitions of sets;
- operations with sets;
- set builder notation;
- principle of inclusion-exclusion;
- the Pigeonhole principle;
- counting simple linear permutations;
- counting simple circular permutations, with and without reflection;
- counting configurations (of grids, dominoes, dice, etc.);
- cryptarithms (arithmetic puzzles) with multiple-digit operations;
- sum of consecutive integers starting at 1;
- other types of sums of arithmetic sequences;
- triangular numbers;
- rules of divisibility;
- remainders;
- prime and composite numbers - the sieve of Erathosthenes;
- factoring into primes;
- rates (direct and inverse relationships);
- conversion of units;
- calculating with fractions;

- simplifying numeric expressions;
- operations with repdigits;
- modeling and solving equations with boxes;
- solving visual puzzles based on: reflections, rotations, or rule-based patterning;

If using our series of workbooks "Competitive Mathematics for Gifted Students", the recommended order of study is:

1. Practice Counting - level 2
2. Practice Arithmetic - level 2
3. Practice Operations - level 2
4. Practice Word Problems - level 2

# Grades 5-6

Pre-requisites: operations with fractions and decimals

- sets of numbers (natural, integer, rational, real);
- interval notation;
- proof that $\sqrt{2}$ is irrational;
- proof that there are infinitely many prime numbers;
- principle of inclusion-exclusion with three sets;
- simplifying expressions with factorials;
- writing fractions as continued fractions;

- non-terminating decimals;
- ratios and proportions;
- divisibility rules;
- bases of numeration;
- least common multiple and greatest common factor;
- perfect powers;
- divisibility properties of some perfect powers;
- factorable Diophantine equations;
- triangular numbers;
- Fibonacci sequences;
- rates;
- mixtures;
- prime factorization of factorials;
- permutations;
- combinations;
- power set of a set;
- Pascal's triangle;
- discrete and geometric probability;
- elements of Euclidean geometry;
- calculate areas by dissection;
- calculate areas by shearing;
- technical drawing skills;
- orthographic projections;

- symmetries;
- simplifying numeric expressions;
- algebraic identities;
- simplifying algebraic expressions;
- simplifying expressions with exponents;
- setting up linear equations and systems of equations;
- inequalities (arithmetic mean - geometric mean, exponential inequalities, triangle);

If using our series of workbooks "Competitive Mathematics for Gifted Students", the recommended order of study is:

1. Practice Word Problems - level 3
2. Practice Arithmetic and Number Theory - level 3
3. Practice Geometry - level 3
4. Practice Combinatorics - level 3
5. Practice Operations and Algebra - level 3

# Grades 7-8

Pre-requisites: algebra I, Euclidean geometry

- invariants;
- summation symbols;
- geometric sequences;

- geometric mean;
- geometric series;
- telescoping sums and products;
- counting the number of positive divisors;
- the sum of positive divisors;
- modular arithmetic;
- linear Diophantine equations;
- higher order non-factorable Diophantine equations;
- systems of Diophantine equations;
- reducibility of rational expressions;
- derive and use Pythagorean triples;
- 2-D and 3-D Euclidean geometry;
- linear functions, equations, and inequalities;
- systems of linear equations;
- algebraic identities;
- simplifying numeric and algebraic expressions;
- word problems with percents;
- rates;
- inverse and direct proportionality;
- word problems with ratios and proportions;
- binomial theorem;
- properties of binomial coefficients;

In our series of workbooks "Competitive Mathematics for Gifted Students" this level is yet to appear. The projected recommended order of study is:

1. Practice Word Problems - level 4
2. Practice Number Theory - level 4
2. Practice Geometry - level 4
4. Practice Counting and Special Topics - level 4
5. Practice Algebra - level 4

# Grades 9-10

Pre-requisites: algebra I, algebra II, Euclidean geometry, trigonometry

- simplifying numeric and algebraic expressions;
- quadratic functions, equations, and inequalitites;
- quadratic functions with parameters;
- exponential functions, equations, and inequalities;
- logarithmic functions, equations, and inequalities;
- exponential growth and decay;
- equations and inequalities with absolute values;
- higher order polynomial analysis (factor theorem, rational root theorem, remainder theorem, Descartes' rule, Viète's relations);
- trigonometric functions;
- trigonometric identities;

- trigonometric equations;
- law of sines and law of cosines;
- solving triangles using trigonometry;
- telescoping trigonometric sums and products;
- analytic geometry (conic sections)
- mass point geometry;
- graph theory;
- advanced counting;
- geometric probability;
- Fermat's little theorem;
- Wilson's theorem;
- inequalities;
- geometric transformations;

In our series of workbooks "Competitive Mathematics for Gifted Students" this level is yet to appear.

At a grade 11 level (AMC 12) specific topics such as: complex numbers, polynomials of complex variable, and cyclotomic equations are added to the AMC 10 topics listed above. At this level, it is the increased difficulty of the problems rather than more advanced theory that is different from the previous level.

Good luck on your problem solving!

www.ingramcontent.com/pod-product-compliance
Lightning Source LLC
Chambersburg PA
CBHW071714040426
42446CB00011B/2053